PENGUIN BOOKS

FIRE AND CIVILIZATION

Johan Goudsblom is Professor of Sociology at the University of
Amsterdam. He has been published widely in various languages. His
works in English include *Dutch Society*, *Sociology in the Balance*, *Nihilism
and Culture* and *Human History and Social Process* (with E. L. Jones and
Stephen Mennell).

JOHAN GOUDSBLOM

FIRE AND CIVILIZATION

PENGUIN BOOKS

To Claartje
and Frank Goudsblom

PENGUIN BOOKS

Published by the Penguin Group
Penguin Books Ltd, 27 Wrights Lane, London W8 5TZ, England
Penguin Books USA Inc., 375 Hudson Street, New York, New York 10014, USA
Penguin Books Australia Ltd, Ringwood, Victoria, Australia
Penguin Books Canada Ltd, 10 Alcorn Avenue, Toronto, Ontario, Canada M4V 3B2
Penguin Books (NZ) Ltd, 182–190 Wairau Road, Auckland 10, New Zealand

Penguin Books Ltd, Registered Offices: Harmondsworth, Middlesex, England

First published by Allen Lane 1992
Published in Penguin Books 1994
1 3 5 7 9 10 8 6 4 2

Printed in England by Clays Ltd, St Ives plc

CONTENTS

Acknowledgements vii

INTRODUCTION:

The Civilizing Process and The Control of Fire *1*

Fire – Civilization – The domestication of fire as a civilizing process – Plan and scope of the book

1. THE ORIGINAL DOMESTICATION OF FIRE *12*

The stage of predominantly passive use of fire – The transition to active use of fire – The formation of the species monopoly

2. THE EFFECTS OF THE USE OF FIRE IN PRE-AGRARIAN SOCIETIES *24*

The widening gap between humans and other animals – Clearing land – Cooking – Warmth, light and other functions

3. FIRE AND AGRARIANIZATION *42*

The second transition – Fire use and agrarianization – Slash and burn: the European case – After slash and burn: increased or decreased productivity?

4. FIRE IN SETTLED AGRARIAN SOCIETIES *55*

Dominant trends – Fire specialists: potters, smiths and warriors – Fire use and fire hazards in cities – Fire in the country

5. FIRE IN ANCIENT ISRAEL 72

Setting and sources – Fire and sacrifice – Fire as a sign of divine power – Fire as a sign of divine anger – Fire in war – Fire in everyday life

6. FIRE IN ANCIENT GREECE AND ROME 95

Setting and sources – Fire in the world of Odysseus: the military regime – Fire in the world of Hesiod: the agrarian regime – The age of the great Greek wars – Fire use and social stratification – Fires and fire-fighting in the Roman world – Fire in religion – Fuel and deforestation

7. FIRE IN PRE-INDUSTRIAL EUROPE 128

The four estates – Fire and religion – Fire in war – Fire in cities – Fire in the country – Fire in technology and science

8. FIRE IN THE INDUSTRIAL AGE 164

Industrialization as a dominant trend – The age of the steam engine and the safety match – New sources of energy: more discrete and diffuse use of fire – Large city fires – Beyond the fire-protected zones: war – Beyond the fire-protected zones: forest fires

9. THE CONTROL OF FIRE AT DIFFERENT LEVELS 194

The individual acquisition of control over fire – Variations in fire use among and within societies – Increased control over fire for humanity as a whole

Notes 216

Bibliography 227

Index 241

ACKNOWLEDGEMENTS

This is intended to be a cool book about fire. However, I would not have been able to write it without the warm support of many persons and of several institutions.

Let me begin by mentioning the institutions: the Department of Sociology and the Postgraduate School of Social Science at the University of Amsterdam, the Netherlands Institute for Advanced Studies (NIAS), Wassenaar, and All Souls College, Oxford. The two university institutes to which I am attached, in addition to providing me with plenty of cause for distraction, gave me the opportunity and the incentive to pursue my work on the rather unconventional subject of the control of fire. At NIAS I found ideal conditions for starting my research; at All Souls for bringing it to a conclusion.

The scope of my subject is vast. I should never have been able to handle it without the help of friends and colleagues from various disciplines. I have profited from the advice and comments of – in indiscriminate Dutch alphabetical order – Anton Blok, Maarten van Bottenburg, D. P. Bosscha Erdbrink, Jan Maarten Bremer, Jan Bremmer, Guy Bush, Han Croon, Joost Crouwel, Sjaak van der Geest, Willy Groenman-van Waateringe, Bart van Heerikhuizen, Judith Herrin, Eric Jones, Bram Kempers, J. P. Kirby, A. Kortlandt, Rita de Koster, Benjo Maso, William McNeill, Fik Meijer, Ravi Mirchandani, Catherine Perlès, H. W. Pleket, Stephen Pyne, Machteld Roede, J. de Roos, Frans Saris, J. M. Schoffeleers, Fred Spier, Jeroen Staring, Margareta Steinby, Mieke van Stigt, Abram de Swaan, Jojada Verrips, Johannes van der Weiden, Nico Wilterdink and Jan Wind. I am particularly grateful to Stephen Mennell and Bryan Wilson, who have read the entire book in manuscript and who have, not for the first time, helped to free my text from Dutchisms. Lesley Levene has given it a discreet and very competent finishing touch.

Some good friends whom I should have liked to thank for encouraging me with their interest in my work have recently died. I cherish their memory: Dick Hillenius, Norbert Elias, Renate Rubinstein.

Only my wife, Maria Goudsblom-Oestreicher, and I know how much this book owes to her. Also on her behalf I dedicate it to our children.

J.G.
Amsterdam,
September 1991

INTRODUCTION:
The Civilizing Process and the
Control of Fire

===

FIRE

The ability to handle fire is a universal human attainment, found in every known society. It is also, to an even greater extent than either language or the use of tools, exclusively human. Rudimentary forms of language and tool use are also found among non-human primates and other animals; but only humans have learned, as part of their culture, to control fire.

According to the simplest definition in modern encyclopedias, fire is a process of combustion, manifested in heat and light. Its immediate effect is destructive. It disintegrates the highly organized structure of organic substances, and reduces them to ashes and smoke. This effect is irreversible; it is impossible for the remains to revert to their original shapes and colours. The phoenix rising from its ashes exists only in the human imagination. Nor does fire have any purpose. The combustion process is blind and aimless; no matter what it touches, if the material is flammable it will be consumed. Of course, the absence of purpose is not peculiar to fire. The same can be said about other natural forces such as rain or wind. But, then, fire has the rare quality of being self-generating. Fire causes heat, and heat in turn causes fire.

Destructive, irreversible, purposeless, self-generating – this does not sound like a very attractive list of properties. What could have induced our ancestors in a distant prehistoric past to tame this wild force of nature and to make it a part of their own societies? What enabled them to do so? And why did they find it worth while? What further consequences did it have, for humanity itself and for its relationship with the rest of nature?

These questions have fascinated people for a long time. There is a rich mythology in which the 'conquest of fire' appears as a great blessing to humankind, often acquired with the aid of a demigod such as Prometheus. In his book *Myths of the Origin of Fire* the British folklorist and anthropologist Sir James Frazer collected a great number of such stories, showing how people all over the world have considered fire as something very precious which somehow fell into their forefathers' possession through subterfuge or good luck. As the French anthropologist Claude Lévi-Strauss has pointed out, a common feature in all such myths is the suggestion that by obtaining fire and being able to cook their food, people became truly 'human'.[1]

In many early myths, fire is treated as if it were a living being, possessed of a spirit with good or evil intentions of its own. Later òn, in the natural philosophies of literate groups in China, India and Greece, fire came to be regarded as one of the primal elements of which the world was composed; according to some ancient cosmologists, it was even *the* major force in the universe. Alchemists and chemists in medieval and early modern Europe continued to put the phenomenon of fire at the centre of their research. In the nineteenth century, however, the notion of fire in the physical sciences was replaced by other concepts such as heat and energy, and it lost its prominent place in scientific theory.[2]

At the same time, fire continued to hold the attention of students of the evolution of human culture. Charles Darwin himself noted in *The Descent of Man*: 'This discovery of fire, probably the greatest ever made by man, excepting language, dates from before the dawn of history.'[3] Many anthropologists in Britain, and even more in Germany, thought along similar lines and wrote extensively on the importance of the mastery of fire for the development of civilization. The British anthropologist Edward B. Tylor, a younger contemporary of Darwin, made an important contribution with a painstaking demonstration that all stories about peoples who allegedly had not mastered the art of controlling fire were false.[4]

In the twentieth century, however, social scientists have tended to follow the example of their colleagues in the natural sciences and have dropped the subject of fire from their agenda. A few anthropologists, such as Omer C. Stewart, continued to draw attention to its importance in shaping human prehistory, and some cultural geographers, most notably Carl Sauer, never ceased to recognize fire as a major agency by which people had changed the face of the earth.[5] Even so, the overwhelming tendency has been to ignore it. Thus the seventeen-volume *International Encyclopedia of the Social Sciences* published in 1968 (the latest edition) does not list the word 'fire' at all, either as an entry or in the general index. It is as if our societies can function without fire, and controlling it poses no problems.

One of the aims of this book, then, is to restore the balance of interest. But drawing attention to fire is not my only purpose. By focusing on the control of fire I wish to raise issues of a more general, theoretical nature as well. The subject can help to remind us how deeply human social life is embedded in ecological processes. And it also shows that these ecological processes have been affected, for a much more lengthy period than is commonly realized, by the activities of humans.

CIVILIZATION

Learning to control fire was, and is, a form of civilization. Because humans have tamed fire and incorporated it into their own societies, these societies have become more complex and they themselves have become more civilized.

This is the basic idea upon which I shall elaborate in the chapters that follow. Implied in it is a concept of 'civilization' that is somewhat at odds with the current usage. It differs both from the way in which the word is most often used nowadays in common parlance and in politics and journalism, and from the more technical meaning that has been given to it in the fields of anthropology and archaeology.

The general function of the word civilization as it is most commonly used in our times was summarized neatly by the sociologist

Norbert Elias in the first pages of his book *The Civilizing Process*:

It sums up everything in which Western society of the last two or three centuries believes itself superior to earlier societies or to 'more primitive' contemporary ones. By this term Western society seeks to describe what constitutes its special character and what it is proud of: the level of *its* technology, the nature of *its* manners, the development of *its* scientific knowledge or view of the world, and much more.[6]

As these words imply, in its common usage the concept of civilization − like the related concept of culture − carries strongly evaluative, ethnocentric overtones. However, it is possible to refine these concepts, and to make them part of a more detached, scholarly discourse. The word culture is now generally accepted in the social sciences as the technical term for referring to all those aspects of behaviour that are 'learned, shared, and transmitted'.[7] I intend to use the word civilization in a similar, non-evaluative fashion.

As a technical term, 'civilization' is nowadays widely used by anthropologists and archaeologists. They give it a rather restricted meaning, however. Archaeologists in particular tend to apply the term civilization exclusively to societies with cities and a system of writing − a type of society that first emerged around 5,000 years ago, some time after the rise of agriculture. This meaning is clearly implied in the titles of such well-known and excellent books as *The Emergence of Civilization* and *Before Civilization* by the British archaeologist Colin Renfrew.[8]

I realize that, as a general rule, it is advisable not to deviate from the standard usage of a concept that has been accepted as a technical term. In this case, however, there are good reasons for doing so. First of all, we have to take into account the lingering evaluative connotations and what they imply. A definition which restricts the concept of civilization to peoples with cities and writing necessarily brings with it the corollary that for by far the greatest part of their history and 'prehistory' people have been *uncivilized*. Most archaeologists avoid stating this conclusion so bluntly, yet it follows logically from their definition.

Another consequence of making a sharp distinction between those peoples who are and those who are not (yet) 'civilized' is that it deprives us of a general term which, in principle, can be applied to the processes of socio-cultural and social-psychological development of people in any society. One may object that we already have such a term – namely, 'culture'. Unfortunately, however, this has strongly static connotations: it refers to *attainments*, rather than to the processes in the course of which these attainments come into being and change.[9]

Thus the American anthropologist Ruth Benedict, in her influential book *Patterns of Culture*, gave a wonderful description of three different cultures and of the profound influence which they exerted over the individuals who grew up in them. However, she passed over altogether the problem of how these three cultures had come to be the way they were; significantly, she took as the motto for her book the words of a chief of the Digger Indians who said: 'In the beginning God gave to every people a cup of clay, and from this they drank their life.'[10] It is in order to stress that we are dealing with processes rather than with unvarying conditions that I have chosen to use the term civilization – as the 'dynamic' counterpart to the strongly static concept of culture. As opposed to the idea of 'culture' and 'cultures' as given structures – which implies a tacit assumption that they have no history or, at the very least, that their history is irrelevant – I shall start with the notion of 'civilization' as a process. And just as the concept of culture cannot be reserved for the so-called 'high cultures', I shall not restrict the term civilization to peoples with cities and writing.

Applying the concept of civilization to all humanity and all human history is not a radical innovation; on the contrary, in doing so I join a long and venerable line of anthropologists and sociologists. However, in recent years this approach has been severely criticized, because originally it often went together with a belief that modern Western culture represented the pinnacle of human civilization to which everyone should necessarily aspire. But recognizing the ethnocentrism in the work of our predecessors need not lead to an uncritical, wholesale dismissal of the task they set themselves: an investigation of the development of human culture and cultures as a coherent process. Indeed, there is a small but growing number of scholars today, representing various disciplines, who see this as a valid and important goal.[11]

In sociology, the most valiant attempt to purge the concept of civilization of its ideological and Europe-centred overtones is still Norbert Elias's study *The Civilizing Process*. The book deals primarily with changes in behaviour among the upper classes in Western Europe between 1300 and 1800. During this period, as Elias points out, standards of behaviour changed profoundly. Far from being haphazard, these changes reflected a thorough transformation of the structure of society. To paraphrase a crucial passage, changes occurred in the social relations between people; consequently, the constraints which people exerted upon each other changed, affecting their behaviour and their emotions, and, thereby, their entire personality structure.[12]

In another passage, Elias noted that it would be impossible to isolate a starting point in the civilizing process.[13] Indeed, at which period in history should we look for the generation that made the transition from being uncivilized to being civilized? How far back do we have to go to meet forebears of whom we could say with confidence that they were in every sense uncivilized, lacking completely in self-restraint learned through constraint exercised by others?

THE DOMESTICATION OF FIRE AS A CIVILIZING PROCESS

The control of fire presents an excellent opportunity for clarifying some of the issues raised by these questions, for it clearly is an element of culture, and as such it has been an integral part of human life for many thousands of generations. The transition from living without fire to living with fire has made life in many ways easier and safer; but it has also created new constraints and risks. The perpetual presence of fire in a human group is a complicating factor; learning to accommodate this complication is a good example of the sort of 'mutation' in conduct which may give fresh impetus to civilizing processes.

Once such socio-cultural mutations have come about, they will not – as in biological evolution – be reproduced automatically. Thus every individual human being has to go through a learning process in order to acquire the skills needed to deal with fire. More generally, in order to become fully human, all people have to pass through a civilizing process of their own, in the course of which they learn, largely from others, how to regulate their own sense impressions and impulses, how to behave and how to think. This we may call the civilizing process at the individual level.

Now, the standards of conduct prevailing in any society at any given time do not have the status of immutable norms. Children in a modern industrial society have experiences with fire, and have to learn skills and habits with regard to fire, different from those of children growing up in a society without matches or lighters but in which fire was commensurately less dangerous. As this example shows, the social norms prevailing in any group at any given time are themselves the result of historical processes. These historical processes represent a second level of civilizing processes: socio-cultural processes – going on in every society – by means of which standards of conduct are handed on from one generation to the next, and in the course of which these standards may also change, whether rapidly or slowly. Elias's study of the civilizing process in Western Europe in the early modern era focused on this second level.

But, as Elias himself was careful to stress, the civilizing process in Western Europe did not start from scratch. No matter how far back we go into the early Middle Ages, nowhere do we find a pristine stage in which people lived entirely without standards of conduct that they themselves as children had learned from their elders and that in turn would be taken over by their own offspring. The European civilizing process clearly formed the continuation, in its own way, of earlier civilizing processes – among the Greeks, the Romans, the Celts, the Germanic peoples, and so on. Nor had any of these earlier societies started from scratch. They all inherited (again, each in its own unique way) more ancient traditions, formed at previous stages. Human history (or prehistory) offers not a single example of an entirely normless group, of a society still completely 'uncivilized'.

We can, then, distinguish a third level at which civilizing processes are to be discerned: the level of human history at large. This third level forms the larger setting within which the civilizing processes at the second (the societal) and the first (the individual) levels take place. At first sight, it may seem forbiddingly vast and complex. However, by tracing the development of the control of fire I hope to show that precisely at this most general level it is possible to perceive some clear overall trends and to distinguish successive stages which form the background to specific historical episodes and individual life histories.

PLAN AND SCOPE OF THE BOOK

In line with the ideas already outlined, I have written this book from a developmental perspective. The frame of reference is human history at large, viewed as a whole composed of the histories of countless specific societies. The underlying plan is chronological, but in some cases, where the features characteristic of a particular stage in socio-cultural development may be observed in different periods, I have deviated from the strictly chronological.

The starting point for my inquiry lies far back in prehistory. How did our earliest human (or hominid) ancestors respond to fire before they could exercise any regular control over it? What enabled them to acquire a certain measure of control? And how did the control of fire, once it was attained, come to be a 'species monopoly', with nothing to equal it among any other living species? I shall discuss these questions in Chapter 1, 'The Original Domestication of Fire'. I have chosen that title – as a variation upon Karl Marx's 'the original accumulation of capital' – in order to indicate that we are dealing with the decisive first stage in a process that is still continuing.

The original domestication of fire formed, I shall argue, the first great ecological transition brought about by humans. In the chapters that follow I shall examine how the relations of humans to fire, and via fire to each other and to other animals, developed thereafter.

This development underwent a radical shift with the emergence of agriculture and animal husbandry, which began some 10,000 years ago and evolved into the second ecological transformation unleashed by humans. The process of agrarianization would have been inconceivable without the control of fire. Once set into motion, it gave new impulses to the civilizing process. The most remarkable new development was, in my view, the progressive differentiation between the ways of life of various groups of people. Increasing cultural diversity, paradoxically speaking, is the common denominator in the pluriform evolution of agrarian societies. The use of fire clearly showed the traces of cultural diversity among and within societies; but it also reflected continuing convergences.

The picture of convergences and divergences in the development of agrarian societies is so multifarious that I shall be able only to sketch its contours. To fill in some details a few examples must suffice. In my choice of these I have followed a familiar trajectory, leading from Mesopotamia (contemporary Iraq, where the first city-states emerged) westward and forward in time, by way of ancient Israel and ancient Greece and Rome, to pre-industrial Europe. This is a rather conventional and 'Europe-centred' course; I have embarked upon it in order not to get lost in an abundance of unordered material. Each of the societies I have selected covered a period of at least 1,000 years, and for none of them do we already have a survey of the various uses of fire. My aim has not been to write an encyclopedic overview; what I have tried to do is to point out some major trends. Further comparative research will be needed to ascertain to what extent these trends correspond to developments in other societies in other parts of the world.

Industrialization formed the third ecological transformation brought about by humans, and again fire played an integral part. The already existing social and cultural differences seemed at first only to be increasing under the impact of industrialization. But there are reasons for arguing that in the contemporary world the tendencies towards convergence are gradually becoming stronger. I will discuss these tendencies in Chapters 8 and 9.

The scope of my inquiry is very large. The action (to speak in terms of classical Greek drama) consists of the human use of fire; the scene is the earth; and the time comprises at least half a million years –

from the very first stages, as reconstructed with great imagination by Jean-Jacques Annaud in his film *Quest for Fire*,[14] to the present, with nuclear physicists able to heat plasma of deuteron and tritium to 150 million degrees. Reading the professional literature – the tops of gigantic icebergs of learning – I have become painfully aware of my own limitations as a non-specialist. At the same time, I have noticed how much room there is for making connections between the various disciplines, and how rewarding is the challenge of looking for these connections.

The subject gives occasion for repeated crossings of the boundaries between different fields. As a sociologist, I make use of the findings of archaeology, anthropology, history, psychology, and even biology and ecology. I have drawn inspiration from the example of other generalists, such as William H. McNeill, who, as a historian, has also explored the ecological dimension of human life, notably in *Plagues and Peoples*, a masterly study of the changing interrelationships between human groups and various microparasites.[15]

As a common frame of reference for the ecological, psychological and sociological aspects of the processes which I am tracing, I have found it useful to bear in mind Norbert Elias's idea of a triad of controls. In each society there is a set of controls extending over non-human events (events, that is, occurring in 'nature' or 'the environment'), over interhuman events (that is, 'social relationships') and over intrahuman events (that is, controls which each individual applies to his or her own impulses and feelings). Our vocabulary for systematically dealing with these interlocking types of control is still very inadequate. The important point is to see them as interrelated and, together, as subject to change.[16]

Equally important is the observation that the triad of controls constitutes, at the same time, a triad of dependencies. Increases in control (which are usually intended) tend to entail increases in dependency (which, by their very nature, are usually unintended). As the human capacity to *control* fire has increased, so has people's inclination to *depend* upon social arrangements guaranteeing its regular availability and minimizing the hazards it involves.

This, then, is the general perspective from which I will approach the control of fire. I will look at the way it has evolved as an integral part of human society. That human beings have learned to handle fire at all I regard as the outcome of socio-cultural development. The possession of fire has made human societies more productive and more formidable, but has also increased their capacity for destruction and made them more vulnerable. As a part of the apparatus by which people control nature, the control of fire has always been and will always continue to be enveloped in social control and self-control. As a subject for investigation, it is fascinating in its own right; at the same time, it can serve as a focus for studying the process of civilization itself.

1. THE ORIGINAL
DOMESTICATION OF FIRE

THE STAGE OF PREDOMINANTLY PASSIVE
USE OF FIRE

Whereas myths typically represent the incorporation of fire into human society as one single event, with a hero playing the central role, it is far more reasonable to think of it as a process extending over many generations, with countless small steps forward as well as periods of stagnation and regression. Even to speak, as is often done, of 'the discovery of fire' may be misleading. As the ecologist Peter D. Moore points out, 'fire has a history on this planet which extends back as far as that of land vegetation itself'.[1] Geological evidence of forest fires is as old as evidence of forest vegetation – about 350 million years. So by the time the first hominids and humans appeared, some 3 to 5 million years ago, fires started by lightning, volcanic eruptions or other natural causes must have been taking place regularly all over the land surface of the earth. Only in areas with very little flammable organic matter, such as the polar regions, deserts and mountain peaks, did fires rarely occur – and these areas were not particularly hospitable to human habitation either.[2] Hominids and humans therefore would not have had to travel far to 'discover' fire; they were just as likely as any other animal to come across a bush fire more than once in a lifetime. And, again like other animals, they would experience such a fire as they would experience rain and snow or heat and cold – as events which came about, over which they had no control, and to which they had to adjust for better or for worse.

They had no means of influencing either the types of fire with which they were confronted or their frequency and duration. The most devastating and frightening were what ecologists nowadays call 'crown fires': fires spreading rapidly through the canopies of trees in woods, attaining very high temperatures and killing most of the vegetation. Such crown fires were not likely to occur frequently, however, for they could come about only if there was a great deal of flammable matter below the canopy to feed them. In most cases any such build-up of dry litter would regularly be destroyed by minor 'surface fires', sweeping with great speed through dry grasses and shrubs, scorching only the tree bases, or by 'ground fires', burning slowly for long periods with the similar effect of eliminating litter.[3]

We may be inclined today to think that the most important thing our early ancestors had to learn was to overcome their fear of fire. However, as the German explorer and anthropologist Karl von den Steinen pointed out as early as 1894, there is no reason to assume that the overwhelming primary reaction to fire was always fear. He described how his servants were in the habit of carelessly abandoning the campfires they had lit, thus sometimes causing big bush fires. The sites of such fires turned out to attract a great many animals.

The fires which we lit during our journey often burned for days and spread spontaneously over large distances. Their influence upon the animal world was curious and striking. All sorts of predators took well-considered advantage of the event; they sought and found their victims, not so much by the bright flames, but rather amidst the smouldering ashes in which many a rodent would lie charring. Numerous falcons were hovering over the dark clouds of the *quiemada*, game was running to it from afar to lick at the salted ashes, preferably at night because they could not hide in the barren plain. The ground radiated a comfortable warmth.[4]

On the basis of these observations von den Steinen continued with a brief digression on the lessons that human beings could at a very early stage have drawn from natural bush fires. At the outbreak of the fire they would have seen game fleeing. Later they would have basked in the glow of the dying embers, and picked up partly charred animals and fruit from the ashes, savouring them. In this manner they would have learned to appreciate the advantages of broiling and roasting, which added not only to the taste of meat but also, more importantly, to its preservation: 'after many days roasted meat which otherwise would have long gone to waste is tasty'.[5]

Von den Steinen clearly set out to shock his modern readers by pointing only to the benefits which fire may have given our early ancestors and leaving aside the dangers almost completely.

Here, however, anyone who is used to looking through the spectacles of [our own] culture will protest. He will note the omission of the horrors which people in prehistoric times experienced from this tremendous phenomenon, but which are little more than the horrors of the scholar whose table lamp may tumble and set fire to his study, his home, and to the entire city with all the valuables it contains. If *I* [he will reason] who have already tamed the power of fire start trembling and shivering as soon as the raging element is set loose, if the gigantic shimmering of flames excites *me* through its impression of fantastic beauty, how must the soul of the poor savage be filled with fright and awe before this mystery.[6]

Even if von den Steinen's words (as in his allusions to a gas or oil lamp) are somewhat dated, the gist of his message still stands. It is quite likely that what today we tend to regard as the only natural human reaction to fire is based to a large extent upon our own experiences with fire in modern society. The ways in which fire manifests itself to people have changed over time, as have the dangers it poses and the fears it arouses. We are nowadays accustomed to strongly regulated and highly inhibited relations with fire – so much so that we may overlook the possibility that fears which strike us as 'natural' and 'rational' may be the result of the very process of domestication of fire. We should be wary, therefore, of projecting our modern feelings about fire on to the attitudes of beings wholly unfamiliar with our way of life, beings with no property to lose, no enduring investments to worry about, but ready to enjoy the warmth, the light, the food and whatever else a smouldering fire might have to offer them.

This is not to say, of course, that they had nothing to fear from fire. The immediate consequences of a fire have always been destructive, then as now. The landscape where a fire has just been raging looks black and barren. Trees are reduced to charred skeletons, plants to ashes. Animals that have not fled in time lie dead amidst the remains, killed by suffocation or desiccation if not by burning. Soon afterwards, however, vegetation will begin to recover and wildlife will return. Most roots will have survived the surface burning and, together with seeds newly blown about or dormant in the soil, rapidly produce new growth. For many plants the longer-term effects of a fire are beneficial; indeed, for some, the so-called 'pyrophytes', they are even essential. Fire destroys their parasites and competitors, and thus has a revitalizing effect on them. Among the pyrophytes are many trees that provide food and shelter to animals, as well as grasses that are eaten by herbivores, the seeds of some of which are also edible to humans.[7]

Not only does a bush fire have favourable consequences in the long run for many species, but certain animals can even profit from it at once. The most immediate advantages, as noted already by von den Steinen, go to birds of prey such as falcons and kites. While the flames are still raging they hover over the fire to hunt fleeing birds and insects. According to an old but persistent legend, the black kite (*Milvus migrans*) may pick up smouldering twigs from a spontaneous bush fire and drop them on dry grass in order to spread the fire and then catch fleeing animals. Experts on birds of prey, however, have been unable to furnish evidence in support of this, or any other legend ascribing to birds the deliberate transport and use of fire. The most likely explanation for these stories is that birds hunting for large insects that have been sent up by the hot air of a fire may occasionally by mistake grasp in their claws a burning twig which they are then forced to drop a little further on in their flight.[8]

When the fire has burned itself out, ground animals may visit the site. First will be predators, come to seek their prey amidst the smouldering remains; later, there will be herds of deer and bovines, venturing near to lick at the salted ashes. Most mammals appear to enjoy the warmth radiated at night by the site of an extinguished fire.

All these responses to fire have been observed frequently in our own time. There is no reason why our hominid ancestors should not have engaged in similar activities centred upon a fire site. Thus the first, and probably by far the longest-lasting, stage in their relationship with fire would have been one marked by the incidental use of natural fire whenever it happened to be available. It has been said that the use of fire at this early stage was 'opportunistic' and not yet 'premeditated'.[9] We might also call it predominantly *passive*. To speak of 'passive use' may sound like a contradiction in terms, but it helps to mark off the initial stage from the next, when hominids began actively to collect and preserve, and later even to make, fire.

THE TRANSITION TO ACTIVE USE OF FIRE

At present any reconstruction of the earliest stages in hominid relations with fire has to be somewhat speculative. We have no ethnographic description of any society lacking the active use of fire. Still, it cannot be denied that humankind must somehow have made the transition from the first stage of passive to the second stage of active use of fire. In the course of this transition, as several commentators have noted, humans changed from an 'ecologically secondary' to an 'ecologically dominant' species.[10] Many other animals recognize the advantages of a fire, and use them when the opportunity offers itself. None of these animals, however, has learned to influence the fire, to keep it burning, to direct its course. Only hominids made the crucial step towards controlling fire to a certain extent and exploiting it wilfully. They learned first, perhaps, to follow fires wherever these occurred, then to preserve a fire for a longer period at its original site, and eventually to transport it to safe and sheltered places which they also made into their own dwellings. When did they accomplish this? What enabled them to do so? And why did no other species attain control over fire?

Although these questions are interrelated, I shall deal with them separately. The problem of when and where hominids or humans first gained control over fire is still unresolved. In the 1940s the South African palaeontologist Raymond Dart claimed that he had found evidence to prove that a primate species akin to *Homo*, but not connected with it through a direct line of descent, was already controlling fire as long as 1,500,000 years ago. Dart named this species, the remains of which were discovered near the South African town of Makapansgat, *Australopithecus prometheus*.[11] Later, however, after critical reviews from colleagues, he withdrew his claim. Then, for several decades, a cave at Zhoukoudiem near Beijing was thought to contain the earliest remains of human-controlled fire; according to general opinion, as long as some 500,000 years ago, humans belonging to the species *Homo erectus* had been tending fires here. In recent years, though, the Zhoukoudiem evidence has also been disputed. Meanwhile, it has been claimed that a few much older sites near Chesowanja (Kenya) and Swartkrans (South Africa) point to human use of fire as early as 1,400,000 to 1,500,000 years ago.[12]

On the basis of the existing evidence I find it difficult to reach my own conclusions, and can note only that the issue is still under discussion. There appears to be a consensus among most archaeologists, however, that sufficiently solid evidence has been found in various parts of Europe and Asia to conclude that *Homo erectus* was using fire at least some 400,000 years ago – that is, long before the appearance of *Homo sapiens*.[13]

While providing the exact chronology remains problematic, we can be reasonably certain about the sequence of stages or phases.[14] First, there would have been a time when no group was in the regular possession of fire; then a time when some groups did, and others did not, have fire; and eventually every human group would have possessed fire. In this section I am concerned with the crucially important second phase in the sequence, the period when some groups began to attain a degree of control over fire.

As long as the archaeological record fails to give conclusive evidence to the contrary, I am inclined to think that this was a very lengthy phase. While it lasted, groups fortunate enough to have found a fire would sometimes manage to keep it burning for months or even years until it was extinguished, either through rain or some other natural cause, or through their own negligence. Throughout this phase, the possession of fire would have remained tenuous and temporary.

What was it that enabled hominids or humans first to enter into this intermediate phase of maintaining a fire for some length of time, and later to be able to hand it down securely from generation to generation? In order to answer this question, it may be useful to distinguish between physical, mental and social preconditions. No doubt such physical features as the erect posture and the concomitant aptitude for carrying objects in the hands and manipulating them were essential. Even in the era of predominantly passive use of fire, one of the advantages which the hominids (and possibly other primates as well) had over other animals was that they were able to handle sticks with which they could rummage in a smouldering fire without getting burned. While poking in the ashes in search of food, they could hardly fail to notice that they might prolong the fire's burning by throwing branches on it. Even more important, however, was the capacity to pick up burning matter and transport it to a place where it would not be extinguished by rain or by wind.

But this was clearly not just a matter of walking upright and having the hands free for carrying something. Fetching branches for a fire implies that the individuals concerned knew what they were doing, and why they were doing it. Keeping a fire going implies foresight and care. Wood had to be gathered, and perhaps even stored during wet periods. Such activities did not come naturally to hominids; they required learning and restraint. Especially when humans began to collect fuel over larger distances, they devoted part of their energy to maintaining something outside themselves, something that was by no means part of their own 'gene pool'. This is not to say that they were acting 'unselfishly'; on the contrary, in caring for the fire they were also caring for themselves. Tending the fire was a form of 'detour behaviour', of 'deferred gratification' like that which was later to become an essential ingredient in agriculture and livestock-raising. Unlike superficially similar complex activities such as nest-building by birds, it was not genetically determined but had to be learned.

The ability to know about fire and the readiness to expend some effort to keep it burning may be regarded as mental or psychological qualities complementing the physical qualities of an erect posture, flexible hands and a large and differentiated brain. However, neither the physical nor the mental abilities would have been of much use to any individual human being had they not been developed in the company of other human beings. Being able to learn from one's elders and being prepared to obey them were additional prerequisites for acquiring a control of fire that, in the succeeding generations, would not be lost again.

In her excellent monograph on fire in prehistory, the French archaeologist Catherine Perlès noted that 'the discovery of the utilization of fire presupposes a *mental* and not a *technical* progress'. Later she added that it also required new forms of social organization.[15] These points are valid. However, rather than setting the technical aspect in opposition to the mental and the social, I think all three should be seen as inextricably linked. Both thought and cooperation were stimulated by the very efforts which the control of fire, as a technical problem, demanded. The technical problem was, at the same time, an intellectual and emotional problem, and a problem of social coordination.

It is unlikely, therefore, that the mental could have developed independently of, or prior to, the other two aspects. Psychoanalytic theory furnishes an example of the extremes to which a one-sided emphasis on the psychological dimension may lead. Sigmund Freud was certainly right when he pointed out that the acquisition of fire demanded the renunciation of certain spontaneous urges. The only urge to which Freud paid any attention, however, was the supposedly irresistible need felt by primal man when he came in contact with fire 'to put it out with a stream of urine'. This infantile desire, connected with the enjoyment of sexual potency in a homosexual competition, had to be overcome.[16]

Of course, for any human group to preserve a fire it is necessary that its male members abstain from peeing it out. But our early ancestors surely had other and more urgent problems to cope with. First of all, they had to prevent the fire going out by itself – either because there was no more fuel for it or because it was extinguished by dampness or rain. In this sense the challenges posed by fire were technical; and the mental and social adjustments that humans successively made were developed to meet these technical challenges.

Clearly the three types of control – over natural events, over social relationships and over individual impulses – were interrelated and mutually reinforcing. As with every human skill, in tending a fire a certain measure of self-control was part and parcel of technical control. The self-restraint necessary for handling the fire competently, without either recklessness or panic, was enhanced by confidence, based upon the observation of others as well as upon one's own experience, that one did indeed control the fire and would be able to use it when needed.

The interrelated technical and mental capacities could be developed and maintained only within a socio-cultural framework. Social co-ordination was needed if only to ensure that there would always be someone to look after the fire. Cultural transmission was needed if the skills, as well as the sense of responsibility and duty associated with the use of fire, were not to be lost. Again, while social co-ordination and cultural transmission were necessary preconditions for the domestication of fire, they were also reinforced by it. Groups in possession of a fire always had to accommodate themselves to the constraints that its maintenance imposed. While subjecting fire to their own purposes, they also had to submit to its requirements. In adapting fire to human needs, they had to adapt human habits to fire. In this sense, the domestication of fire also involved 'self-domestication' or 'civilization'.

THE FORMATION OF THE SPECIES MONOPOLY

Reflections on the preconditions necessary for the control of fire by our early ancestors are at best informed conjectures. There is no possibility of putting the interpretations I set out above to the test. I do hold, however, that they are plausible and make sense. They are not at odds with any known facts and they throw some light upon the general question of how the transition to the active use of fire could have come about. As I hope I have shown, the ability to control fire was made possible by the concurrent development of specific social, mental and physical capabilities.

A problem that still needs further consideration is how the ability to control fire came to be an exclusively and universally human characteristic. Maybe the solution has already been given in the discussion of preconditions. It is possible that the combination of social, mental and physical properties needed for lasting control of fire emerged only in the course of human evolution, and that no other species possessed them. If this was indeed the case, then there is no problem. However, it is also possible to see the issue in a very different light. There are, in fact, two mutually exclusive ways of looking at the preconditions for the control of fire. On the one hand, we can treat the requirements of care, caution and foresight as so forbiddingly difficult that it puts them beyond the reach of any species but *Homo*. But we can also take a broader view, one that takes account of the fact that other primates living today, chimpanzees in particular, come remarkably close to possessing the configuration of social, mental and physical traits needed for the active use of fire. Ethologists have done very little research on this aspect of behaviour (surprisingly little, I think), but there is at least some evidence to suggest that chimpanzees may indeed be capable of keeping a fire burning for some length of time.[17] If this is so, it is likely that *Australopithecus* and other higher primates that are now extinct had evolved even further in this direction. The question then arises, what prevented them from continuing on this route?

In answer I can put forward only a hypothetical reconstruction of what may have happened. My scenario is based upon the idea that elimination contests and processes of monopoly formation, such as have been observed taking place *within* human societies, may also have occurred at a much earlier stage *between* groups of hominids and other animals. The exclusive control of fire by humans, in other words, may be viewed as the outcome of *inter*species struggles, comparable to the *intra*species wars out of which, at a much later stage in history, the state monopolies of organized violence and taxation emerged.[18]

Initially fires could be used only at the original site of a wildfire. Hominids, gathering around such a site, could train themselves and each other in some rudimentary techniques: throwing various objects into the fire to see if they would burn, or lighting the far end of a stick and swinging it around – maybe to intimidate others. As the advantages of fire came to be more clearly perceived, access to such sites would have become increasingly desired and contested. Hominids already close to the fire might employ burning sticks to ward off invaders. In such struggles, an erect posture and manual dexterity were obvious advantages, and so – to at least the same extent – were an increased ability to communicate, group cohesion and discipline.

Again, the triad of controls was self-reinforcing. Groups already near to a fire were always at an advantage. If they were capable of developing a proper balance of daring and caution with regard to the fire, and if each generation succeeded in transmitting this acquired characteristic to their young, they would in the long run become increasingly proficient in handling fire, and, consequently, more formidable as opponents in violent struggles.

Thus the balance of power shifted. Groups having a high level of control over fire could effectively keep others, with a lower level of control, away from it. The others inevitably lost the means of training themselves and their young in the art of handling fire, so that whatever skill their forebears might have developed in this respect was lost. Again, we should not conceive of this overall trend as being made up of events all leading in the same direction, but rather as a protracted and complicated process, in the course of which the pendulum may have swung many times at different places. All we know is the outcome: human monopoly over the control of fire.

The idea that other primates may at some time have known rudimentary forms of active fire-use fits in with the theory of the Dutch ethologist Adriaan Kortlandt that the living great apes represent the descendants of more humanlike ancestors who were driven from the savannahs into the forests by proto-hominid competitors and were thus forced to adopt the lifestyle of tree-dwellers. Kortlandt speaks in this context of a process of 'dehumanization'. Indeed, nothing in the physical constitution of chimpanzees would prohibit them from throwing wood into a fire or lighting sticks in it. One possible handicap for apes in coping with fire might have been that sparks could not be flicked off from their fur as easily as from the thinly haired human skin. This difference between humans and apes –

which is not mentioned by Desmond Morris in his book *The Naked Ape* – may have worked to the advantage of humans, although on the other hand, greasy fur, such as some primates possess, may also confer protection against injuries from fire and against the pain of a superficial burn. All in all, in the greater scheme of physical, psychological and social characteristics needed for the control of fire, hair growth is not likely to have been the decisive factor.[19]

What can be stated with confidence is that the human monopoly of the active use of fire was established a very long time ago. All other animals are defenceless against it. They have learned to perceive it as part of the apparatus by which humans dominate and, as the biologists S. L. Washburn and C. S. Lancaster put it, 'terrorize'[20] their world, and from which they can only flee. In some species, exclusion from the use of fire may have blocked further socio-cultural development; at the same time, the exigencies of living *with* fire may well have contributed to the singular development of the human capacity for language and thought. Thus on both sides of the figuration the human monopoly of fire contributed to widening the gap in power and behaviour between humans and all other animals.

Finally, there is the problem of explaining how the ability to control fire became not just uniquely but also universally human. Once again, I can suggest only a hypothetical scenario. Once the struggles for fire were under way, and certain hominid groups were gaining a decisive advantage, neighbouring groups could not afford to lag behind. They would either have to become equally competent in handling fire or eventually suffer the fate of the vanquished: submission and accommodation, flight or extinction. In the long run, no human groups without fire survived. The transitional stage, in which some groups did and others did not possess fire, came to an end. The control of fire had become, as it is today, both an exclusive and a universal attribute of human societies.

2. THE EFFECTS OF THE USE OF FIRE IN PRE-AGRARIAN SOCIETIES

====

THE WIDENING GAP BETWEEN HUMANS AND OTHER ANIMALS

The control of fire could not fail to have far-reaching consequences. By its very nature it affected the relationships between human beings and the world they lived in, including their relationships with other animals. It also affected the social relationships among and within human groups. And its consequences inevitably extended to the way human individuals learned to perceive the world and to modify their own behaviour. It is difficult to keep these three aspects – the ecological, the sociological and the psychological – apart, since they are so closely interrelated. But it is also important to bear the distinction in mind, because it will enable us to perceive the connections clearly.

In the relationships with the surrounding world the first great change to occur – setting the stage for a trend that has been gaining momentum ever since – was an increasing differentiation in power and conduct between hominids and other large mammals. Especially with regard to those primates that were most closely related to humans, the formation of the species monopoly did much to boost the process of differentiation.

In our own time, human supremacy has become so firmly established that people have created nature reserves and zoos in which the anthropoid apes, along with other animals, enjoy special protection against the dangers by which they are threatened – especially other people. Occasionally, apes living in captivity are given access to some form of domesticated fire. Thus in the 1950s several chimpanzees in the Johannesburg Zoo were taught to smoke; and not only did they become addicted to cigarettes but they also learned to light them and to extinguish the stubs. Watching their behaviour the palaeontologist A. S. Brink was deeply impressed by 'the bodily skill, visual acuity, manual dexterity and mental ingenuity' they displayed.[1]

Yet no matter how deft and clever the chimpanzees observed by Brink were in manipulating a burning cigarette, there is no evidence whatsoever for chimpanzees in their natural environment displaying active control over fire. Even if they have any inborn potential capacity for controlling fire, this has not been developed spontaneously in any known chimpanzee group. We need not rule out the possibility that the ancestors of the contemporary chimpanzee and other primates living a million years or more ago sometimes managed to keep a fire burning for some time. However, since the original domestication of fire began, the ability of hominids to control fire has developed further and further, whereas among the other primates any tendency in this direction was arrested. Some primates, like *Australopithecus*, became extinct; others, like the chimpanzee, survived but stayed far behind the hominids in their ascent to ecological dominance.

Again in South Africa, the archaeologist C. K. Brain found dramatic evidence for the shift in the balance of power between hominids and large predators. A cave at Sterkfontein was dominated for countless generations by big cats who drew australopithecene victims into its dark recesses for consumption. At a later stage the tables were turned and a new group of hominids (identified as *Homo erectus*) made its appearance; they 'not only had evicted the predators, but had taken up residence in the very chamber where their ancestors had been eaten'. How they managed to do this is not recorded, but, as Brain comments,

it could surely have been achieved only through increasing intelligence reflected in developing technology. It is tempting to suggest that the mastery of fire had already been acquired and that this, together with the development of crude weapons, tipped the balance of power in their favour. The tipping of this balance represented, I think, a crucial step in the progressive manipulation of nature that has been so characteristic of the subsequent course of human affairs. It was a step the robust australopithecenes apparently failed to make, and their extinction was doubtless hastened by predators they were powerless to control.[2]

Inspired by the work of Brain and his colleagues, the British writer Bruce Chatwin drew a vivid and persuasive, if necessarily speculative, picture of one of the first steps in humanity's ascent to hegemony. He evokes the scene of hominids as the favourite prey of big cats (*Dinofelis* or the 'false sabre-toothed tiger') prowling about at night which had acquired, from generation to generation, a special taste for human flesh. Overcoming this lethal danger, according to Chatwin, must have been the greatest victory in human history. It could be achieved only through cooperative effort; and the weapon, still according to Chatwin, could only have been fire.[3]

In reading this scenario it is important to remember that the ordinary sabre-toothed tiger (a formidable animal, considerably larger than the Indian tiger living today) was still living in America only some 15,000 years ago. Yet even if Chatwin's reconstruction is somewhat over-imaginative, the fact remains that the most pervasive trend in the development of human society over many thousands of generations has been the increasing differentiation in behaviour and power between human groups and all other mammals. Both sides in the figuration were affected by shifts in the balance of power: as human hegemony spread, life chances for humans became greater and human populations grew in number and extended their territories, whereas other animal species either became extinct or had to adjust their way of life to the new situation.

In a popular book on human prehistory, published in 1965, the American archaeologist F. Clark Howell provided a powerful description of a site near Torralba in Spain, where the skeletons of numerous large mammals were found, including almost thirty elephants. The presence of stone tools among the remains pointed to human activity; apparently this had been a hunting site of *Homo erectus* some 400,000 years ago. What made Torralba especially interesting were traces of fire in the vicinity, suggesting that the human hunters had first driven the animals to a precipice by setting fire to the surrounding grassland, and then killed them after they had jumped forward into an adjacent bog. As Howell himself put it:

There was also a quantity of material that shows different degrees of burning ... These materials were not so concentrated in any one place as to suggest the presence of continuous fires over a long period of time. Rather they were thinly and very widely scattered. Whoever had been lighting these fires was apparently burning grass and brush over large areas. This evidence, plus that of the elephant bones concentrated in what was once a bog, suggests that the setting of these fires had been purposeful – to drive the unwieldy elephants into the mud.[4]

This suggestion, supported by colourful drawings, has elicited a great deal of comment. Thus, typically, the biologist Melvin Konner has interpreted the scene as follows: 'whole herds of elephants were apparently driven by grass fires to their deaths, stampeded over a cliff, in much the way recent inhabitants of the American Great Plains stampeded bison.'[5] Before going along with this interpretation it is worth noting, first of all, that Howell himself, in a more seriously scientific publication about his research at Torralba, makes hardly any mention of the massive killings with the aid of fire. And even more sobering is a recent analysis of the evidence by the American archaeologist Lewis Binford, according to whom there is no conclusive evidence for game drives and mass kills at Torralba: the charcoal deposits may have been caused by natural fires, and the accumulation of bones could just as well point to human scavenging as to hunting.[6]

Lack of certainty about what precisely happened at Torralba should not blind us, however, to the overall trend. Human groups developed more effective hunting methods, and became increasingly formidable as opponents to all other animals, whether they were predators preying on human flesh, predators competing with human hunters for their quarry, scavengers competing with hunters, or herbivores which themselves were a quarry for the human hunters. Testimony from witnesses to tell us how *Homo erectus* groups stalked their prey is an impossibility, but there is plenty of circumstantial evidence suggesting that fire was an invaluable weapon. Even into our own time – long after the invention of the bow and arrow by *Homo sapiens* some 20,000 years ago greatly contributed to the effectiveness of human hunters and reduced their need to rely on fire[7] – fire continued to be used by individual hunters to smoke a single small mammal out of its shelter, or by well-organized groups to drive a whole herd of elephants into an ambush. Ethnographic reports provide many interesting instances. Thus, when a wandering group of !Kung Bushmen in the Kalahari desert robbed some lions of their prey and the lions came back at night, they would not venture within the circle of light cast by the campfire.[8] Today human hegemony has become so firmly established that, even in small agrarian communities, guarding domestic animals or scaring away wild animals from the crops is generally regarded as a 'particularly light task' which is usually left to children or old people.[9]

CLEARING LAND

In using fire for hunting, human groups changed the land they inhabited – at first perhaps inadvertently, and later deliberately. The most drastic transformations were to occur after the emergence of agriculture and modern industry. Yet, as early as the stage when human groups subsisted wholly by gathering and hunting, they were already making a strong imprint on the landscape. Their chief agent in doing so was fire.

Fire is self-generating: its heat causes further material to ignite. As long as fuel and air are present, the fire will continue to burn and spread. Once human groups came to possess fire, therefore, they were in a position not only to prolong the burning of a single fire but also to 'reproduce' it, to use it for making other fires. In a way, the very word fire is ambiguous and somewhat misleading, since it may refer both to *a* fire – a singular, isolated event – and to fire in general – a process that occurs over and over again. The point is that with *a* fire people could continually make *more* fires. Their control of fire enabled them to proliferate fire.

And this is what they did. Going from place to place, they would light a campfire wherever they decided to stay. When leaving they would hardly bother to put it out; as the anthropologist Omer C. Stewart has noted, 'before the manufacture of fire was discovered ... there must have been much more thought and energy given to preserving fire than to extinguishing it'.[10] Thus the frequency of bush fires and savannah fires inevitably increased, as abandoned human fires were added to those brought about by natural causes.

Moreover, gatherers and hunters also lit fires intentionally. A primary reason for doing this was to drive game animals out of their lairs in the bush. The fire might also chase away predators and snakes, while the combination of smoke and fire might kill insects and small parasites. After the fire was over, nuts and fruits that had lain hidden in the undergrowth could be spotted and picked up more easily. And there were still more advantages to be gained from a fire. One immediately obvious result of clearing the undergrowth would have been to allow human hunters to see further and move more easily. In the longer term there might have been even more beneficial effects. A fire would create favourable conditions for particular plants, especially grasses and legumes which need direct sunlight. Fire, as humans came to recognize, could be employed to stimulate this kind of vegetation, which in turn would attract game.

Thus there was a whole sequence of events likely to occur after a piece of land had been exposed to fire. Having learned to recognize this sequence, humans could then go further and bring it about intentionally, thereby creating ecological circumstances beneficial to themselves.

Besides these 'material' benefits, they probably also derived certain emotional pleasures from their burning practices. Setting fire to a tract of land could be a way of 'appropriating' the land, of making it their own and establishing 'dominion' over it. Of course, there is nothing in the archaeological record to confirm such a motive, but anthropological observations among foraging peoples in our own time strongly suggest that it has played a part.[11]

Initially, hominid or human groups must have been small and widely scattered. Their burning practices, therefore, would not have had massive impact on the landscape. Still, human groups equipped with fire and the propensity to use it on a large scale have been around for many thousands of generations. As the American geographer Carl Sauer has remarked: 'Wherever primitive man has had the opportunity to turn fire loose on a land, he seems to have done so from time immemorial.' The net result was, as Sauer and many others after him have pointed out, that long before the introduction of agriculture, the vegetation in many parts of the world had already been strongly affected by human interference with the aid of fire.[12]

Naturally most of the ecological effects of early human burning practices were obliterated by the drastic changes in climate when, during the late Pleistocene, the last interglacial period was followed by a long ice age, reaching its peak between 22,000 and 16,000 years ago. Any lasting effects of the burning practices of gatherers and hunters still visible today date from the latest period in the earth's geological history, the Holocene, which started when the last glaciation came to an end some 10,000 years ago.[13]

Soon after the last ice-caps had receded and trees could begin to grow again in what were now the temperate zones, humans began applying their ancient burning practices to the new virgin woods. Archaeologists have found evidence of such incipient deforestation as a consequence of fires lit by humans in many different regions, including England.[14] As the human population increased, the 'fire economy' was intensified and eventually led to the emergence of agriculture – a development I shall discuss at greater length in Chapter 3.

In some parts of the world the transition to agriculture did not take place until quite recently. Most of the land in North America was not used for crop cultivation when the first European colonists arrived in the seventeenth century. The early settlers in New England found an open and park-like landscape, with very little undergrowth. The prevalence of this type of vegetation was mainly due to the practices of the Indians, who would let a fire blaze over the forest soil twice every year. The fire, as one colonist wrote, 'consumes all the underwood and rubbish which otherwise would overgrow the country, making it unpassable, and spoil their much affected hunting'. Another colonist wrote, in much the same vein, that 'this burning of the Wood to them they count a Benefit, both for destroying of vermin, and keeping downe the Weeds and thickets'. The historian William Cronon, from whose book *Changes in the Land* these quotations are taken, concludes:

Indian burning promoted the increase of exactly those species whose abundance so impressed English colonists: elk, deer, beaver, hare, porcupine, turkey, quail, ruffed grouse, and so on. When these populations increased, so did the carnivorous eagles, hawks, lynxes, foxes, and wolves. In short, Indians who hunted game animals were not just taking the 'unplanted bounties of nature'; in an important sense, they were harvesting a foodstuff which they had consciously been instrumental in creating.[15]

When later generations of European colonists began to penetrate into the Midwest of North America, they came across a very different landscape: the prairies. The endless open spaces, inhabited by large herds of buffalo, seemed to them to be primeval. However, the nature of this landscape was largely the result of the burning practices of Indian hunters, who had been systematically burning down tracts of forest in order to create more grassland for buffaloes and other herbivores. As the historian Stephen Pyne observes in his book *Fire in America*, 'except for the High Plains, where the short grass expanses were more or less determined by climate, nearly all these grasslands were created by man, the product of deliberate, routine firing'.[16] Whenever, in our own time, a piece of prairieland was left to itself instead of being subjected to regular burning and grazing, the grasses were soon overgrown by trees and the prairie spontaneously reconverted to forest.

A similar situation prevailed in Australia. When in 1644 the Dutch sailor Abel Tasman set eyes on the western coastline of the continent, he saw 'fire and smoke ... all along the coast'. Voyagers into the interior later noted 'the very extraordinary devastation by fire' which had affected the vegetation everywhere they went. Sylvia Hallam, who has collected many such observations from nineteenth-century travelogues, concludes that at that time 'burning, though the work of a comparatively small population, was impressive in scale, frequency, and undoubtedly in vegetational effects'.[17]

No doubt natural fires, caused by lightning, occurred regularly all over the continent. Their frequency and effect were greatly increased, however, by the deliberate burning practices of the Aborigine population. Leaving aside the question of whether or not these practices should be referred to as 'firestick *farming*',[18] we may note that there is abundant evidence to suggest that well into the nineteenth century Aborigines regularly set fire to the land in order to hunt, to keep the country open and to stimulate the growth of fresh grass. English travellers who first saw them at work with their firesticks were astonished by 'the dexterity with which they manage so proverbially dangerous an agent as fire'. As one observer wrote in 1831:

At this season they procure the greatest abundance of game ... by setting fire to the underwood and grass which, being dry, is rapidly burnt ... With a kind of torch made of the dry leaves of the grass tree they set fire to the sides of the cover by which the game is enclosed ... The hunters concealed stand in the paths most frequented by the animals and with facility spear them as they pass by. On these occasions vast numbers of animals are destroyed. The violence of the fire is frequently very great and extends over many miles of country; but this is generally guarded against by burning it in consecutive portions.[19]

As Sylvia Hallam's material clearly shows, the effects of the Aborigines' firing were not merely ecological, affecting the flora and fauna, but also sociological and psychological. Setting fire to a tract of land was an investment, entitling people to certain rights of usage and creating emotional ties to it.

Hallam is careful to note that the situation encountered and documented by nineteenth-century Europeans must be seen in its historical context. It is likely, as she observes, that human groups had used and valued fire ever since they first arrived in Australia some 30,000 (or perhaps even more) years ago. Throughout that period conditions must have changed, however, as the human population increased and as more land came to be effectively exploited. As Hallam suggests, 'initial relatively random firing with severe effects would lead to more regular firing of vegetation increasingly adapted to increasingly frequent and regular burning'. In other words:

The type of regime we have seen clearly portrayed in nineteenth-century sources and implied back to the seventeenth century had developed over a span, not of two, but of more than 200 centuries. There is no need to envisage a highly regulated, seasonal, sequential firing within defined territories as spanning the entire time range. Such a close framework of responsibilities need have developed only as population densities became relatively high, number of groups per unit area high, and group range correspondingly restricted.[20]

COOKING

The process of combustion that we call fire is in itself a blind natural process. We are used to calling it destructive, because it reduces highly organized matter to a lower state of organization or integration, and it does so in a manner that is irreversible. Sometimes, however, the primarily destructive effects of a fire are conducive to processes of reorganization and reintegration at a higher level. This happens in ecological systems when a fire burns away dead matter and fungi, thus giving new scope for plant and animal life. This is also what has happened in the domestication process, in the course of which humans have come to some extent to control the energies released by combustion and to use these in the organization of their social life.

When gatherers and hunters set fire to bushes in order to create more favourable grazing grounds for game, they acted on this principle. They used the immediately 'destructive' effects of fire, anticipating that after a while these would lead to 'productive' results. The principle that was applied here at the ecosystem level could also work on a much smaller scale: in cooking. The destructive force of fire was set to work in order to produce food more edible and tasty, and better suited to human consumption, than the ingredients were in their original, raw condition.

The initial use of fire for cooking, as for hunting and clearing land, must have been incidental and mostly passive. In our own time, it has been observed that after a bush fire chimpanzees carefully went over the charred ground under afzelia trees in search of beans. When raw, these beans would be too hard to bite, but roasted by the fire they could be crumbled easily between the chimps' molars. Stella Brewer, who reported this, also noted that for over a week large fallen trees continued to smoulder slowly to ashes. It made her think about what a short step it was 'from searching for the comparatively few cooked seeds to gathering the hard raw seeds and deliberately cooking them in these natural ovens that were scattered about in the valley'.[21]

As with other uses of fire, it is impossible to tell when, and how often, this 'short step' was made. But then again, there can be no doubt that its consequences were far-reaching. In the long run, cooking brought about physiological as well as psychological and sociological changes.

The physiological implications of the transition to eating cooked food are complex and not yet altogether clear, as has been shown in a recent survey of the literature by Ann Brower Stahl.[22] Since the regular consumption of a wide range of cooked food (and not just incidentally provided seeds) most probably influenced the human digestive system in the long run, it cannot be inferred with certainty from present knowledge what the hominid diet must have been like before the domestication of fire. Some highly plausible conjectures can be made, however. By letting the fire do some 'pre-digestive' work, cooking opened up new food resources. Most importantly, a variety of vegetable substances, especially leaves and legumes, were made suitable for human consumption once fire had destroyed toxic substances and tough fibres. Thus the range of plant food was extended considerably and many new sources of protein, starch and carbohydrate became available. In addition, exposure to fire might protect various sorts of food, meat in particular, against decay, so that they could be preserved for a longer time. It seems safe to conclude, therefore, on purely nutritional grounds that, since it facilitated biting, chewing and digesting, and since it removed or lessened the harmful effects of noxious compounds, cooking enhanced human life.

As well as having these nutritional effects, cooking also affected social organization and mentality. What is true of control of fire in general applies even more so to cooking: it became exclusively *and* universally a human skill, requiring not only certain biogenetic preconditions but social organization and cultural transmission as well. The 'short step' noted by Stella Brewer eventually became an enormous stride as cooking developed into a complex set of activities far removed from the simple reflex chain of hunger, food-seeking and eating. In this respect, as the French archaeologist Catherine Perlès remarks, 'the culinary act' brought about a divide between 'the animal world and the human world'.[23]

The higher productivity yielded by cooking could be attained through the same forms of cooperation and division of labour that had enabled groups to control fire. While some were guarding the fire, others would take part in a communal hunt or go out by themselves to collect fruits and tubers. It was mostly men, and maybe also semi-adult boys and girls, who did the big-game hunting; gathering vegetable food and cooking were largely the work of women. But no matter how and by whom the food was obtained, cooking formed an intermediate stage before it was consumed; and different individuals might be involved in obtaining, preparing and consuming it.[24]

Like the control of fire, cooking is an element of culture. It has to be learned, and this learning is done in groups. It demands a certain amount of division of labour and mutual cooperation, and also, at the individual level, attention and patience. One has to watch the food from time to time and to postpone eating it until it is well cooked and has cooled off a little. According to some authors, the attention people had to pay to their cooking would have supplied them with 'the first subtle and intimate knowledge of matter', thus forming the basis for the further development of the empirical natural sciences.[25] And already at a much earlier stage, the social coordination and the individual discipline acquired as a result of cooking might have useful spin-off effects in other activities as well.

Several psychoanalytic authors have suggested that even the very act of eating was deeply affected by cooking. The softening of fibres by the fire would have made it possible to reduce the part played by the teeth in tearing tough substances, and to rely increasingly on the far more versatile hands.[26] Pulling food apart with the hands, before the mouth touches it, is generally regarded as a more 'human' way of eating than using the hands only to hold the food fast while the teeth and lips do all the cutting and tearing. The shift from 'oral' to 'manual' may even be regarded as a primary spurt in the civilizing of eating habits, long before the introduction of chopsticks or knives and forks.

The increasing use of one's hands, and, even more, of utensils designed to take over part of the work of the hands, made eating a more differentiated activity. Of course, when we say that people eat with chopsticks, we imply that they use their hands and mouths as well; but the expression clearly indicates where the social priorities lie. In the long run, the possibility of eating cooked food, composed of various ingredients and consumed with manual skills and perhaps even with special utensils, has added to the differentiation of diets and eating habits among humankind, and, consequently, to the gamut of grounds for social distinction. Anthropology abounds with examples not only of variety in diet and ways of eating, but also of the revulsion evoked by other people's eating habits. The Algonquin Indians in north-east America used to call their more northerly neighbours 'Eskimos', which meant 'raw-meat eaters'.[27] This name, which has become familiar in all Western European languages, is by no means the only case of one group stigmatizing another as eaters of raw meat. In societies with greater internal differentiation and a more involved hierarchy, cooking and eating habits have continued to be among the aspects of behaviour most commonly used by members of the different social strata as a basis for mutual identification and invidious distinctions.

I cannot hope to have covered in these few pages all the implications of cooking. The main categories, however, have been mentioned. First, there were the physiological consequences: the possibilities of extending the diet with substances which in the raw state would lend themselves to human consumption only with difficulty if at all. No less important were the social and psychological consequences, as cooking led human groups to cultivate eating habits of their own, by which they could distinguish themselves both from all other animals and also from each other. Distinctions first developed in interspecies relationships were carried over into intraspecies relationships.

WARMTH, LIGHT AND OTHER FUNCTIONS

Clearing land and cooking represent, so to speak, two prototypical forms of fire use, requiring different kinds of fire. In clearing land, the fire was 'let loose' to spread quickly over a large area. In cooking, it was kept within bounds to produce even and concentrated heat.

Of these two major forms of controlled fire, clearly the kind used for cooking was the more permanent. As such it could serve a variety of other functions too. As a source of heat and light it gave protection against cold and darkness. It kept predators and other animals at bay. Because of the comfort and security it offered, it could be a focus of group life and enhance communication and solidarity. It was also useful for a variety of practical purposes, such as splitting stones or sharpening wooden tools. Embers from the cooking fire could be taken to other places, and the smoke too could be put to good use – repelling insects, driving out game from its shelters, and sending signals over large distances.

This brief and incomplete list of fire's uses may be read as a supplement to my discussion of the preconditions of the control of fire in Chapter 1. If we wish to comprehend the course of the domestication process, it is not enough to enumerate the traits that hominids had to develop to enable them to acquire some control over fire. We also have to consider what made it worth while for them actually to exploit and develop this potential.

Fire, to repeat, is a physico-chemical reaction which, under varying circumstances, will have a number of consequences, both in the short and in the long run. Humans have learned to perceive and appreciate certain consequences and to bring them about by creating the appropriate circumstances. Thus they could assure themselves of certain desired results. In the language of systems theory, we might say that they produced 'effects with a positive feedback'; we might also speak of 'post-conditions' or, more commonly, of functions.[28] The point is that only the continuous pursuit of the advantages to be gained from having a fire could have kept the domestication process going over so many thousands of generations.

But then again, this was not the whole story. Whether people were aware of it or not, their regular use of fire inevitably led to more consequences than those they directly pursued. One of these unforeseen consequences was the formation, in the long run, of a new 'ecological regime', a fire regime that became an integral part of human society. When people used fire to clear land or to ward off other animals, they imposed this socio-cultural regime upon their natural environment. When they spent part of their energies looking after the fire or collecting and storing fuel, they subjected themselves to the fire regime.

The fire regime made human groups stronger as 'survival units' *vis-à-vis* both hominid and non-hominid competitors.[29] In the long run, only those groups which were equipped with fire and prepared to live under a fire regime survived. Adding the force of fire to their own, they could make their societies more productive and more formidable. The increases in productivity, achieved by more effective hunting as well as by cooking, may initially not have been great. Yet in time they could not fail to result in a rise in the level of material comfort and an increase in human numbers – or in what modern economic historians would call, respectively, 'intensive growth' and 'extensive growth'.[30]

The use of fire as a source of heat and light undeniably brought about a rise in the standard of living (to apply another concept from contemporary economics to ancient prehistory). We still tend to associate warmth with comfort, and light with lucidity and splendour. To be able to produce warmth and light at will, throughout the year, must indeed have made for a great improvement in living conditions. Even if the possession of a fire was not absolutely necessary for people to endure the cold and damp winters of northern Eurasia, it certainly made them more bearable.

In this way the control of fire also facilitated territorial expansion and, consequently, population increase or 'extensive growth'. At a number of sites there is evidence of human habitation in cold regions some 400,000 years ago without any trace of fire. Of course, ashes from these sites may have been blown or washed away, but it is also possible that groups there managed to survive without fire. From later periods, however, there is evidence only of groups with fire.[31]

Just as the warmth radiated by fire furthered the expansion of the human dominion territorially, the light it spread offered temporal expansion, making it possible to fill dark evenings with work, play and ritual. Of course, even with fire humans remained highly dependent upon the cycles of the seasons and of day and night. However, their dependence became less direct; fire served as a buffer against the extremes of cold and darkness, creating small enclaves of warmth and light.

Again, it was not fire alone that made life in a cold climate more endurable. As the examples of the Eskimos or the inhabitants of the high Andes in modern times show, shelter and clothing and a diet rich in fat may offer sufficient protection.[32] But fire certainly helped in bringing about a more agreeable micro-climate. This was all the more advantageous to human groups, since the colder temperate zones in Eurasia offered them what was in many ways a better environment than the tropical or semi-tropical parts of Africa from where they originated. In the north, as William McNeill points out, there were far fewer micro-parasites to pester and to kill them.[33]

All these advantages combined, in the long run, to make human groups increasingly more dependent upon fire. Like certain plants that can thrive only in an environment regularly visited by fire, they became pyrophytes, addicted to fire. And thus, by implication, they had no choice but to continue living under a fire regime.

As a socio-cultural regime, the fire regime tended to strengthen group ties. It was by virtue of their membership of a group that people enjoyed the benefits of having a fire. The extent to which, from early on, a group may have been identified with its fire can be inferred from the custom, noted among gatherers and hunters in our own times, of referring to one's group affiliation by saying to 'which fire' one belonged.[34]

Clearly, the fire furnished a setting around which a group could gather and engage in common activities well into the night. Conceivably, after the species monopoly had been established, it also helped to create a bond between groups, which would then 'lend' fire to each other in case of need. There is no empirical evidence to support this idea, but it offers an alternative to Lévi-Strauss's hypothesis that the earliest bond between groups was based on the exchange of women.[35]

To its many benefits, the fire regime added one more pleasure: that of having power over the fire, and, through the fire, over animals and plants. There is evidence from recent times that pride and joy could also be derived from exposing people to terrible ordeals by fire, as happened in initiation rites or in the torture of prisoners or outcasts.[36]

Needless to say, the power exercised through fire was essentially social – that is, it could be sustained only by a group. And while groups might differ in many respects, if only because of their varying habitats, they tended to be remarkably similar in their fire regime. The basic principles of fire control, as discovered by hominids, left little leeway for variations, and led to more or less the same adaptations all over the world, regardless of climate and geography.[37] It was simply impossible to keep a fire burning for long without at least some social cooperation and division of labour in order to guard it and fuel it. As people came to rely more and more upon the various functions of fire, they also became increasingly vulnerable to its loss. This might be another reason why fire was surrounded with reverence and ritual. At the same time, because of its destructive force, fire never ceased to be potentially harmful, and handling it always required discipline.

Since no hominid or human group could afford to start from scratch and invent for each new generation new solutions to the problems posed by keeping a fire, every group had to rely largely upon socially standardized procedures, or rites. Rites, as learned alternatives to 'instincts', have continued to be attached to the use of fire in every known society to the present day.[38] As it is most likely that this has been the case ever since hominids first learned to keep a fire going, it stands to reason that the control of fire has also contributed to the development of the general human capacity to engage in ritual.

Rites always involve both prescriptions and prohibitions. According to the French philosopher of science Gaston Bachelard, for children 'fire is initially the object of a general prohibition', and 'the social interdiction is our first general knowledge of fire'.[39] These remarks are at once both highly perceptive and yet limited in scope. They pertain primarily to children living in houses in an agrarian or industrial society. The situation is different in a society of gatherers and hunters who, as adults, regularly set the bush ablaze in order to clear the land. A child growing up in such a society need not receive such severe warnings and might even be encouraged to play with fire in order to learn how to handle it.[40] But even in societies that are less haunted by fear of fire than ours, every child has to find out about the cultural restrictions pertaining to domesticated fire and has to learn that it is dangerous to ignore them.

Because of the discipline it inevitably required, the domestication of fire was also a civilizing process, involving the development of social codes in accordance with which individuals had to behave. This is not to say that controlling fire made human beings more loving and peaceful; on the contrary, one of the long-term concomitants of the civilizing process was an increased capacity among humankind to turn the destructive power of fire against anything deemed undesirable and fit for devastation. The point is that handling fire or explosives, even for purely destructive purposes, never ceased to demand great caution. As people succeeded in stoking increasing numbers of larger and hotter fires, they needed tighter regulation of their social relations and individual impulses in order to keep those many fires under control.

3. Fire and Agrarianization

THE SECOND TRANSITION

Long after they had domesticated fire, human groups began to extend their care and control over selected animals and plants. This marked the beginning of the second great ecological transformation brought about by humans. It could not have taken place without the firmly established control of fire. Once it was under way, it in turn deeply affected the further development of the part played by fire in human society.

The initial transition from gathering and hunting to agriculture and animal husbandry was not necessarily abrupt. A group that started to cultivate a few crops would not have to give up its older ways altogether. If any, there would have been only very few agrarian societies from which gathering and hunting disappeared completely. However, the proportion of products acquired in the older way inevitably diminished as agriculture and animal husbandry advanced.

For many thousands of generations, the two processes of rising standards of living (or 'intensive growth') and increases in human numbers (or 'extensive growth') were almost imperceptibly slow. We can now see that both began to accelerate towards the end of the last glacial period, some 20,000 to 30,000 years (no more than 1,000 generations) ago. I mention just two indications: first, pointing to intensive growth in particular, was the appearance of rock paintings in the interior of caves; and second, pointing to extensive growth in particular, was the spread of the human population over every continent, including the new worlds of the Americas and Australia.

Both intensive and extensive growth were speeded up even more with the emergence of agriculture and animal husbandry, some 10,000 years (or 300 to 400 generations) ago. With this second ecological transformation humanity once more entered a new stage in its history. Yet there were also remarkable similarities and continuities.

The questions I have asked with regard to the domestication of fire are also relevant to the domestication of plants and animals: what were the pre-conditions underlying the process, and what were the post-conditions or functions that kept it going and made it so compelling that agrarianization turned into a dominant trend? It has been estimated that 10,000 years ago the human population lived exclusively by gathering and hunting; that 500 years ago their numbers had been reduced to 1 per cent; and that in 1965 the figure was less than 0.01 per cent.[1] In retrospect, the past 10,000 years appear to have been a period of transition between two stages: a preceding stage in which there were no societies with agriculture, and a consecutive stage (which we have entered in the twentieth century) in which there are no longer any societies without it.

The domestication of plants and animals was in several significant ways comparable to the domestication of fire. It too involved the move to a more active and regular use of natural resources. Groups of people 'tamed' originally 'wild' forces of nature and learned to tend, guard and exploit further these forces within their own human domain. After incorporating fire, they now incorporated certain selected plants and animals into their own societies. They extended their care and control over these species, and initiated a process of artificial selection to improve their qualities for meeting certain human needs. Just as for millennia they had supplied their fire with fuel, and protected it against wind and rain, they now began to feed and cultivate their domesticates, and to protect them against competing species and parasites.[2]

In this way, human communities created high concentrations of plants and animals which could provide them with useful products – most importantly food. In the long run, this increase in the productivity per unit of land led to considerable increases in human numbers, and to a potential rise in the standard of living, although, for reasons to be discussed later, the latter was in many societies attained only by a minority of the population.

At the same time, as societies became more populous and, in part at least, more prosperous by virtue of agriculture, they also became more dependent on this form of production. Like the domestication of fire, the control gained by the domestication of plants and animals brought with it increased dependence, on that which was controlled, as well as on the technical and organizational apparatus which made such control possible. In order to comprehend the basic structure and dynamics of agrarian societies, it is necessary to bear in mind this twofold tendency. On the one hand, by deliberately extending their mastery over non-human resources, these societies became increasingly more productive and more formidable. However, the very same advances in control also had the effect of increasing overall capacity for destruction and of rendering these societies more vulnerable to various kinds of catastrophe.[3]

FIRE USE AND AGRARIANIZATION

It is almost impossible to imagine how people could have begun to practise agriculture if they had not been thoroughly familiar with the art of handling fire. For one thing, they needed a fire with which to cook. The first crops cultivated on any large scale were cereals belonging to the plant order of grasses. Owing to their high nutritional value and their ability to withstand storage for long periods, cereal grains formed a very appropriate staple food for a human community; to serve this purpose, however, they had to be made more easily digestible with the help of fire.

A second and very different reason why the control of fire was a precondition for the domestication of animals and plants was the human predominance over all other mammals, which was grounded partly upon the use of fire. The human species' monopoly over fire was so solidly established by that time, and is today so taken for granted, that it is seldom given separate attention in this context. Yet it deserves mention. Their generally recognized hegemony in the animal kingdom enabled people not only to bring certain species, such as goats and sheep, under direct control but also – and this was at least as important – kept most of the remaining 'wild' animals at a distance from their crops and their herds.

It is very likely that experience in controlling fire furthered plant and animal domestication in yet another way. Like the domestication of fire, the domestication of plants and animals implied an extension of the human domain. A new kind of ecological regime was established, an agrarian regime, imposing new demands and constraints upon both the physical environment and the human community itself. Cultivating crops and raising livestock were, like tending a fire, forms of 'detour behaviour' in which people cared for something not of their own kind; this detour behaviour was not innate but had to be acquired through social learning. The long familiarity with a fire regime probably helped to prepare people for the strains of an agrarian regime full of self-imposed renunciation for the sake of 'deferred gratification'.

The most striking, and in the literature the most frequently discussed, relation between the use of fire and the emergence of agriculture is the ancient tradition of burning off land with an eye to food production. As noted in Chapter 2, burning the land was a widespread practice among foraging peoples, who would in this way create better conditions for both gathering and hunting. These firing practices were probably intensified at the end of the last ice age, some 12,000 to 10,000 years ago, when an increasingly serious ecological crisis confronted people in various regions.[4] By this time the human population had spread over all continents; even the Americas and Australia, although sparsely populated, were inhabited from coast to coast. Advances in hunting techniques had very likely already caused a considerable thinning out of the megafauna. Under these conditions, the rise in temperature must have brought about a most precarious situation precisely in those areas which, as a result of intensified gathering and hunting, were the most densely populated, such as the 'Fertile Crescent', stretching from Mesopotamia, through Asia Minor, to Egypt. As the temperature increased, ice-caps began to melt, causing an eventual rise in the sea level of almost 130 metres. Fertile delta regions on the coasts were thus lost to human habitation. At the same time, the tree line shifted and savannah-like areas to which people were well adapted were overgrown by forests offering far less favourable conditions to gatherers and hunters.

It seems very probable that the advance of the forests stimulated people to intensify their firing practices. As the American anthropologist Henry T. Lewis has pointed out, people some 10,000 years ago were capable of a highly judicious use of fire; they could employ it in such a manner as to further a type of vegetation which provided large quantities of food for themselves and for the animals which were their prey. Whereas frequent fires would reduce the total biomass of vegetation, they stimulated lush young plant life with a high nutritional value for animals, thus increasing the animal biomass well beyond that found in dense forests. When, in response to the changing ecological conditions, people intensified their burning practices, they continued to add new experience to the ancient 'folk science of fire'. With the help of this knowledge, they were able to encourage the growth of a vegetation dominated by grasses, which provided not only edible seeds for themselves but also stalks and seeds which, although indigestible to humans, could be eaten by sheep, goats, gazelles and other beasts of prey. This burning-economy, directed at increased productivity in hunting and gathering, was, as Lewis remarks, an important 'preadaptation to agriculture', no less important than the development of stone-grinding tools and storage facilities.[5]

In a similar vein, the British archaeologist Paul Mellars has argued that increasingly efficient use of fire for hunting prepared the way for pastoralism.[6] By prudent burning Neolithic hunters were able to create areas of ground with a vegetation cover that would attract large numbers of herbivores while at the same time offering high mobility and easy visibility to the human hunters. These conditions permitted the formation of greater concentrations of people, subsisting on the highly specialized hunting of large game. When subsequently, as a result of the very efficiency of hunting, game threatened to become less abundant, people learned to become more selective in the choice of animals for slaughter. What may at first have been a luxury – greater discretion in either killing or sparing specific animals – gradually developed into a vital necessity for the survival of both the human predators and the herds upon which they preyed.

Continuing along the lines set out by Lewis and Mellars, the Italian historian of agriculture Gaetano Forni has argued that the Neolithic hunters who burned the brushwood to further the growth of tender grasses and shoots which would attract herbivorous game were already acting as 'breeders of animals'. Similarly, gatherers who burned the woodland to promote the growth of grasses instead of trees were not essentially different from 'cultivators'.[7] The same argument has been used by several authors who claim that the Australian Aborigines practised a form of proto-agriculture known as 'firestick farming'.[8] In a way these labels are, of course, merely a matter of definition. It seems to me, however, that a significant distinction is lost when the stage of intensified gathering and hunting with the help of fire is already characterized as agriculture. I think it is important to see that burning practices were, in many cases, a precondition for the emergence of agriculture and should not be identified with it.

As the case of the Aborigines shows, the transition from foraging with the aid of fire to agriculture was not universal. In the long run, however, agrarianization did turn out to be the dominant trend, and societies without agriculture have by now virtually disapppeared. The fact that this dominant trend began some 10,000 years ago in the Fertile Crescent of the Middle East (and perhaps in some other pats of the world as well) can probably be explained as a reaction to the 'food crisis' sketched above. More sophisticated burning practices did not automatically lead to the rise of agriculture, but they contributed strongly to it.

SLASH AND BURN: THE EUROPEAN CASE

Even with favourable climate and soil conditions, the seeds of cereal grains and other grasses would be able to germinate only if the ground was not already covered by other plants. For agriculture and pastoralism to develop it was therefore necessary for pioneer farmers to clear away, as far as possible, any already existing vegetation. The most efficient means for doing this was fire.

In the majority of cases the land which people brought under cultivation as arable or pasture was originally covered by forest. To burn away at will a tract of forest was not a simple matter. At most times of the year, trees and shrubs contain a large amount of sap and do not burn easily; at other times, the vegetation may be so dry that once a fire breaks out it will be extremely difficult to contain. In order to accommodate these difficulties people have developed the practice known as 'slash and burn' or 'swidden agriculture'.

Slash and burn is yet another example of how the problems associated with the use of fire gave rise to solutions that are remarkably similar all over the world. In spite of numerous regional variations, with a plethora of distinct indigenous names, the method of slash and burn amounts everywhere to the application of the elementary principle that dead wood is much more flammable than live wood.[9] Therefore, in an area marked for clearance, people begin by killing the trees, cutting off the branches and ring-barking the trunks. These activities are usually carried out at the beginning of a dry season. When, some months later, the dead litter is set ablaze, the flames can quickly destroy the wood.

In societies with a more intensified form of agriculture, both land and wood have become increasingly scarce and, consequently, the practice of slash and burn is regarded as primitive and wasteful – as under modern conditions it frequently is.[10] Thus, when Charles Dickens, on his visit to the United States in 1842, saw the practice still being used in the Allegheny mountains, he found it 'sad and oppressive' to come upon great tracts where settlers had been burning down the woods, and he almost showed pity for the trees whose 'wounded bodies lay about, like those of murdered creatures, while here and there some charred and blackened giant reared aloft two withered arms, and seemed to call down curses on his foes'.[11]

When slash and burn was first conceived, however, it represented an important step in the civilizing process of humankind – a major breakthrough in which the immediately destructive effects of fire were utilized in a longer-term ecological strategy. This strategy required elaborate technical and social skills. It presupposed not only the availability of strong and sharp axes but also the capacity to plan ahead for at least several months. The entire procedure comprised a series of successive steps: first, selecting an appropriate plot of land; then, carrying out the various preliminary slashing operations; and, after a considerable interval, choosing the right moment to set the dry wood on fire. The timing of the actual burning required particular knowledge and care. As has been observed in Africa in our own time, if a tract of forest is set on fire too early, 'much of the precious ash will be blown away and lost; if it is left too late and the rains have come, the burn will be incomplete and the crop poor'.[12]

Experiments imitating prehistoric conditions have confirmed that the method of slash and burn, if carried out with patience and skill, could result in very fertile soil for cultivating crops.[13] Almost unavoidably, other plants would thrive in this soil as well. In the absence of ploughs, the competing plants, the 'weeds', would usually take over completely after a few harvests. As long as there was enough suitable wild terrain in the surrounding area – either unworked 'primary' or previously cleared 'secondary' forest – the simplest solution was to clear an adjacent section by slashing and burning and to abandon the fields that were in use. A cycle could thus develop in which farmers prepared a new section of land every few years, leaving behind the fields they had been cultivating, only to return some years later when forest vegetation had regenerated. Under optimal conditions a community might profit from this cycle for many generations.

These optimal conditions would include, in the first place, a natural environment that was sufficiently resistant to erosion or other forms of depletion. On inclines and in arid regions the chances were that after a fire much of the loose topsoil containing the fertile ashes would be washed or blown away. Moreover, the population had to be not too numerous, and not susceptible to rapid growth, so that the land could remain fallow for a long enough period. And then people needed to have the prudence to abstain from 'prodigal or inept agricultural practices which sacrifice future prospects to present convenience'.[14]

The way in which the first clearing of land for agriculture took place in different countries and regions is still a matter of research and debate among archaeologists. However, as early as 1952 the British archaeologist J. G. D. Clark drew a general picture of the development in Europe which can be regarded as paradigmatic in several ways for other parts of the world as well. In his book *Prehistoric Europe: The Economic Basis* Clark described how the frontier between forest and open fields shifted in a north-westerly direction from Asia Minor via the Balkan peninsula and the Danube basin, a process which must have started in the sixth millennium BC and has continued on into the modern era. Before the advent of agriculture, a dense deciduous forest (formed after the last ice age) stretched almost unbroken over all of central and northern Europe. The very domain of farming, as Clark notes, had to be carved out of this primeval forest. Under these conditions, pioneer agriculture was extensive.

Patches of forest would be cleared, sown, cropped and after a season or two allowed to revert to the wild, while the farmers took in another tract. In this process burning played a vital role, since it converted timber into ash and so provided a potash dressing for the virgin soil. So long as the forest lasted and until clearance outstripped the capacity of the woodlands to replace themselves, the system of *Brandwirtschaft* or burning-economy was capable of supporting the prehistoric peasants at a very tolerable level of well-being.[15]

Clark's words seem to anticipate the picture that anthropologists have subsequently drawn of an 'affluent society' in which many Neolithic slash and burn farmers lived.[16] However, the idyll, if indeed there was one, did not last. Along with the crops, other plants that flourish in the open – including various species of grasses, heathers and ferns – made their appearance. Heavy grazing by sheep and goats had the effect, more often than not, of leaving strips of land that had been cleared by burning devoid of forest growth, even if they were no longer cultivated. This, together with increases in the human population, resulted in a steady process of deforestation.

According to Clark, these environmental changes inevitably led to social changes. In many areas the slash and burn culture reached a point of exhaustion, after which continuation of the same old ways was no longer possible. Life became harsher and more violent; instead of 'peaceful peasants' came 'warriors'.

Surely one is dealing here with the effects upon human history of an immense ecological change wrought unthinkingly by the Neolithic farmers and their livestock. The crisis when it came extended far beyond the sphere of animals and plants and involved not merely the economic basis but the whole outlook of large segments of the population of prehistoric Europe. In many parts at least the fat times of forest farming were over for good and all. The stored-up fertility of the virgin soil had been taken and the potash from the burnt woodlands had been absorbed.[17]

After the slash and burn economy had passed its prime, the most important means for working the land in Europe became the plough. Up until recently, though, there was still a frontier of agriculture at the far north and east, in Finland and Russia, where slash and burn continued to be practised.[18] Around the Mediterranean, by contrast, the period of slash and burn was over long ago. Some archaeologists have argued that there is precious little evidence pointing to the use of slash and burn in Neolithic Europe.[19] The empirical basis upon which they rest their case, however, seems to relate to a period when farmers were no longer following this practice. Cultivating the land by plough, which became the dominant practice all over Europe, represented a phase following, and made possible by, the period in which the land was originally cleared by fire.

The scenario sketched by Clark is, of course, very general and would have to be modified to fit particular cases. But as a theoretical model it seems valuable. The succession of stages can be seen in many other historical eras, and in geographical locations as far apart as India and Peru. In a great variety of settings the same ecological sequence has occurred, and, concomitantly, the same sociological process of increasing social inequality, social tension, violence and warfare.[20]

AFTER SLASH AND BURN: INCREASED OR
DECREASED PRODUCTIVITY?

As people became more numerous and arable land became scarcer, food production was intensified, and farmers began to suppress the natural vegetation by means other than fire. New methods of working the ground were discovered, such as irrigation, ploughing and the use of manure, which made it possible to use an area, once it had been cleared, year after year without interruption. The role played by fire in preparing the land was reduced to such minor operations as burning off stubble or refuse.

There is a clear and easily comprehensible order in the sequence of change. First, enclaves of arable land, created by means of slash and burn, were used on an impermanent basis in a system of 'shifting cultivation'. Then, gradually, larger and more lasting settlements were founded and 'swidden agriculture' was relegated to the fringes of the agrarian world. As more people came to live in villages and towns, their use of fire became both more specialized and more highly regulated. Specialization led to the construction of different types of hearth, oven and lamp for a variety of domestic and professional purposes. Apart from these, however, people tried to make their villages and towns into 'fire-protected zones' as much as possible. Fire was thus brought even more tightly under human control. At the same time, it became more dangerous, because of the proliferation of ovens and hearths and also the concentration of flammable material possessions which people accumulated in these fire-free zones.

The extensive firing of forests and bushes – so characteristic of the preceding stage – became a thing of the past in societies with more intensive methods of food production. An interesting question, one that has a direct bearing upon my argument, is whether the later methods of working the land – irrigation, ploughing and manuring – were more or less productive than the earlier cultivation by fire.

In a study that has gained the status of a modern classic in its field, the economist Ester Boserup has given an unequivocal answer to this question: productivity with slash and burn methods was higher than all subsequent forms of cultivation. She came to this surprising and provocative conclusion by defining productivity as 'yield per man-hour'. According to this definition, slash and burn methods are indeed highly productive, for, as Boserup rightly observes, 'the fire does most of the work'.[21]

However, while slash and burn may have demanded relatively little labour, it did require a great deal of land. An ecological and socio-logical approach to the question of productivity therefore leads us to a conclusion very different from Boserup's strictly economic approach. What counts is no longer the average yield per hour of an abstract individual, the fictitious *Homo economicus*, but the total yield actually produced on a particular plot of land by and for a particular human community.

When they engage in agriculture, people concentrate the crops that they find useful in one area and fight off the growth of other plants. If successful, their labour will make the soil produce a higher yield of the desired products. We are thus dealing with a real pro-duction increase by the addition of labour. Defining productivity by the average individual output per hour, on the other hand, may give us a convenient statistical measure but it obscures our picture of what was really taking place when agriculture became more labour-intensive. We fail to see that even if many people had to work harder and had to live more monotonous lives under less healthy conditions, the societies of which they were a part were becoming more pro-ductive.

The transition from slash and burn to more labour-intensive methods of land use nearly always implied an increase in productivity. In most cases the increased productivity was expressed directly in population increase or 'extensive growth'. This is not to say, of course, that the relationship was monocausal. Rather, demographic growth, intensification of labour and increase in productivity were mutually reinforcing processes. Once a population was growing in size, occasioned by more labour-intensive methods of land use, it would need even more labour in order to sustain the next generation – unless alternative methods of production or exploitation were found. In this way peoples who practised labour-intensive forms of agriculture were, so to speak, 'doomed' (or, as the Book of Genesis put it, 'cursed') to till the land from generation to generation.

As suggested by J. G. D. Clark, the pressure to work was less strong during the first phase of agrarianization, when there was still sufficient land to be cleared by slash and burn. But it shrank inevitably as the population increased and less land had to be cultivated by more intensive labour. This may help to explain to some extent why the very productivity of advanced agrarian societies seemed to condemn the majority of their members to an existence full of hardship into which they were born and to which they had to adjust – often under priestly exhortations that this was the will of God.

At the same time, the high level of productivity, attained by hard work, also made these societies more formidable than those whose subsistence was based on foraging or on shifting cultivation. This point may suggest an answer to the question of how the intensification of agriculture came to be an irresistibly dominant trend, in the course of which societies of settled farmers and peasants generally superseded those of gatherers and hunters and of shifting cultivators. It becomes exceedingly difficult to comprehend the compelling dynamics of this development if, by defining productivity in merely individual terms, we deny the real increase in collective productivity in advanced agrarian societies.

4. Fire in Settled Agrarian Societies

===

DOMINANT TRENDS

The emergence of agriculture and the raising of livestock ushered in a new era in human history. From now on, fire was no longer the only non-human source of energy that had been brought under human control. Gradually it ceased to be the prominent focus of group life that it had been for many thousands of generations, becoming increasingly more dispersed over various sorts of specific 'containers', such as hearths, ovens and lamps, beyond which it was not allowed to spread. At the same time, its use was subjected to stricter regulation. As more and more people came to live in villages and towns, they tried to make these settlements into 'fire-protected zones' where the use of fire was permitted only within specifically designated confines. (Like so many social rules, this one applied only in times of peace; during war those same zones that people ordinarily tried to keep protected from fire became, for their enemies, the targets for arson.)

From now on farmers had to till the same fields year after year, using such labour-intensive methods as irrigation, ploughing, terracing and fertilizing the soil with manure from domesticated animals. Only at the outskirts of settled agrarian society was there still sufficient land available for slash and burn. As contemporary anthropologists such as Marshall Sahlins and Marvin Harris have pointed out, the hard labour that went into food production did not make life healthier or more agreeable for the majority of the people.[1] Nevertheless, intensification of agriculture was the dominant trend. At a much earlier stage, human groups with fire had survived, whereas groups without fire had not. Now, for a period of at least 5,000 years, there was a similarly dominant tendency, not only for groups with agriculture to supersede groups without it, but also for groups with intensive agriculture to supersede groups with extensive agriculture.

All this implied further differentiation. The previous era had been marked by the increasing differentiation in behaviour and power between human groups and other animals. The agrarian era, too, was marked by increasing differentiation, again in both behaviour and power; not just *vis-à-vis* other animals, however, but also, and especially, among and within human societies. The following extract may serve to illustrate this point.

This is how the present life of man on earth, King, appears to me in comparison with that time which is unknown to us. You are sitting feasting with your ealdormen and thegns in winter time; the fire is burning on the hearth in the middle of the hall and all inside is warm, while outside the wintry storms of rain and snow are raging; and a sparrow flies swiftly through the hall. It enters in at one door and quickly flies out through the other. For the few moments it is inside, the storm and wintry tempest cannot touch it, but after the briefest moment of calm, it flits from your sight, out of the wintry storm and into it again. So this life of man appears but for a moment; what follows, or indeed what went before, we know not at all.

This passage is taken from the *Ecclesiastical History of the English People* by the Venerable Bede, completed in 731.[2] The speaker is a high-ranking priest who is urging his king to embrace Christianity. Apart from its intrinsic value as a parable, the passage is also of historical and sociological interest. The scene, an assembly of noblemen, comfortably seated around the hearth in the hall of their king, well insulated from the cold outside, could just as easily have been taken from Homer's *Odyssey*. It would have fitted the circumstances of the warrior élite in the Aegean Bronze Age 2,000 years earlier equally well as those of the Anglo-Saxon noblemen.

Moreover, within the societies evoked by these authors, the warriors and their strongholds formed only one part in a much larger configuration. Besides them, there were other classes of people, living in other surroundings, and having other relationships to fire. Thus, to mention only the major classes, in many agrarian societies there was also a class of priests which sometimes even claimed priority over the warriors, entitling themselves the 'highest caste' or the 'first estate'; there was a class of artisans, craftspeople and traders; and there were the peasants, who continued to form the bulk of the population.

We are faced, then, with the seemingly paradoxical situation that differentiation among and within agrarian societies was a common trend in agrarian society at large. This, it seems to me, was indeed a major socio-cultural trend throughout the 10,000 years of 'agrarianization', when agriculture and the raising of livestock became the dominant means of subsistence. Differences among societies increased, first, as some groups gradually switched to agrarian production while others continued to live by foraging, and, subsequently, as some agrarian societies came to rely on increasingly more intensive methods of cultivation whereas others continued to practise slash and burn. Within settled agrarian societies, great differences in behaviour and power developed between the various social classes.

Looking back from our present vantage point on 10,000 years of agrarianization, we can see converging as well as diverging trends in the process of socio-cultural development or 'civilization'. In comparison with the long-lasting pre-agrarian phase, the divergences stand out as the most striking. The human civilizing process now proceeded in distinctly different ways – first of all, in different societies in various parts of the world, and, secondly but no less importantly, among different social strata in each of these societies. The very processes of cultural divergence and social differentiation were common structural features of all advanced agrarian societies. It is against this background that, on the one hand, highly distinct forms of civilization evolved in China, India, Persia, Egypt, Rome, Mexico, Peru, and that, on the other hand, in all these different societies highly similar systems of social stratification emerged, all characterized by sharp contrasts in lifestyle and power between the ruling élites and the mass of peasants and landless poor.

To summarize the interplay of both converging and diverging tendencies, we can say that intensification of agriculture was generally accompanied by a further increase in the numbers of people, by increasing concentration of people in permanent settlements, by increasing specialization of people, by increasing organization of people in larger economic, religious and political units, and by increasing social stratification, or a division of people into upper and lower tiers with greater or lesser access to power, property and prestige.

Fire had become so much a part of society that each of these trends also affected the control of fire. As societies became more populous, the number of fires increased as well. These fires were increasingly concentrated within the 'fire-protected' zones of towns and villages, in hearths, braziers, ovens and lamps. The use of fire was also increasingly more specialized, with various crafts and professions developing their own skill in employing it. With proliferation, concentration and specialization came new forms of organization, if only to meet the growing need for fuel. And unavoidably, the use of fire also reflected the process of stratification, as some people came to command enormous supplies of fuel, and could use fire for impressive displays of power, whereas others might never call any fire their own.

It may be helpful to keep the complex of interrelated trends in mind when looking in greater detail at instances in the development of the control of fire in advanced agrarian societies. While in individual societies and in separate periods these five tendencies were at times interrupted and even reversed, they persisted, in the long run, for humanity at large. And it is also worth noting that whenever one of the five trends went into reverse, the other trends changed direction too.

FIRE SPECIALISTS: POTTERS, SMITHS AND WARRIORS

A Fatal Configuration

The five trends listed above formed a single cluster, each inconceivable without the others. Thus specialization could not have proceeded without organization, stratification could not have proceeded without organization, and so on. The whole cluster of trends combined to make agrarian societies more complex, and to give rise to different classes of people with different social functions, such as peasants, craftspeople, warriors and priests.

For peasants, fire continued to be indispensable for a wide range of purposes, from domestic cooking to the burning of stubble and waste. Apart from seasonal fire festivals, the uses of fire tended to be routine, and people learned the skills needed to handle a fire while still children. In some areas peasants were, in the long run, confronted with a depletion of fuel resources, caused either by their own consumption or – as will be shown by examples in later chapters – by wood supplies to cities or the incursions of highly fuel-intensive industries such as mining.

While the peasants' use of fire was primarily practical, its ceremonial functions predominated for the priests. Being less strongly bound by practical restrictions, the priests could develop a wider range of cultural variations in fire use. Thus highly specific fire rituals were developed in Hinduism and Zoroastrianism.[3] Another well-known expression of such relative autonomy in the formation of cultural traditions with fire are the funeral rites still surviving on the island of Bali, where great towers are erected for the cremation of deceased members of the nobility. In Chapter 5 I consider the way priests in ancient Israel might have used fire rituals as a means of establishing and reinforcing collective religious identity.

Here I will concentrate on two groups of specialists who were fire masters *par excellence*: the potters and the smiths. Both used the destructive power of fire in order to produce socially valuable objects. An important group of the objects manufactured by smiths comprised weapons, designed to kill. In order to understand the social position of the smiths (and the potters as well) it will therefore be necessary to consider also their peculiar relation to the warriors, who were to become their most powerful clients and patrons.

It was the social fate of both potters and smiths that, like the peasants, their occupations made them extremely vulnerable to organized violence and deprived them of the time and the means to defend themselves against it. According to the British anthropologist Ernest Gellner, 'agrarian society is doomed to violence. It stores valuable concentrations of wealth, which must needs be defended, and the distribution of which has to be enforced.'[4] The potters and the smiths had no more power than the peasants to resist this tendency, and to prevent agrarian society from turning into military–agrarian society dominated by warriors.[5] Willingly or unwillingly, they contributed to this trend, and they were engulfed by it.

Potters

Creating objects out of clay or metal represented, by its very nature, almost the paragon of productive work. People could perform this work only with the aid of fire. They could let the fire destroy the original compounds in which clay or metal was found, and thus bring about new substances with a new shape.

The creation of pottery basically depended on two operations: modelling a piece of clay into the desired shape, and heating it in such a way that it would harden irreversibly, becoming incapable of entering into a chemical combination with water again. For a very long time, people must have witnessed the process of clay hardening fortuitously each time they used it for a hearth. The oldest known remains of intentionally baked clay, in the form of figurines, date from the Upper Palaeolithic (between 20,000 and 30,000 years ago) and have been found in various places on the Eurasian continent.[6]

Baking vessels of earthenware became practical only after people had adopted a sedentary life. Initially, perhaps, the production was carried out by itinerant specialists travelling with their tools from village to village, and firing their pots on open fires. The pots as such, however, were too heavy and too fragile to be moved around. They were suitable only for people living in permanent settlements.

In such settlements, potters could build ovens, enabling them to control the heat more effectively than was possible on an open fire. Clearly, building an oven and using it to good effect required many skills. Testing the quality of the clay to be used, cleaning, moulding, drying, firing and cooling it, and finally decorating the finished object – all these activities demanded knowledge, skill, attention and patience on the part of the individual craftsperson. By implication, they also involved social and cultural preconditions. The oven, besides being a technical accomplishment, also represented a capital investment. To employ it, the potters needed a regular supply of clay and fuel. They could practise their craft only if they were able to count on a reasonable measure of security from raids. This may help to explain why many of them were, at first, attached to a temple or a palace.

Earthenware was useful for storing wheat, nuts, oil and other foodstuffs. It contributed greatly to a community's productive capacities by making its products more durable. Food and seeds could be preserved and made inaccessible to animals over a great length of time. As people had to rely on one and the same harvest for longer periods, their fates became progressively tied up with that harvest. Not only was it now possible to preserve food in pots but doing so became imperative for survival. The fragile earthenware, containing vital necessities of life, symbolized both the enhanced productivity of sedentary peoples and also their vulnerability, especially as the military specialists were able, with the aid of metallurgy, to increase their powers of destruction.

Smiths and Warriors

Metal ores were originally used in much the same way as stones: as cutting tools, as objects of decoration and exchange, or, in a pulverized form, as a pigment (the famous rock paintings in the caves of Lascaux and Altamira were coloured by red ochre and other ores). Actual metallurgy does not appear to have begun until the fifth millennium BC, with artefacts made of smelted lead.[7]

Current opinion about the details of its development in various parts of the world has been summarized by the American archaeologist James Muhly in the statement that 'The idea of a unified sequence of steps or stages in technological progress ... seems to be gone forever.'[8]

Yet when we look at human history as a whole, a common

structure can still be discerned among the many local and regional variations. As Muhly also stresses, the invention of metallurgy rested on the dramatic discovery that a hard, intractable rock could be turned into a pliable and malleable metal. There must have been a phase, somewhere,

when all the critical first steps were taken in learning what metal was, how metals behaved, how metal had to be worked (in ways quite different from the familiar techniques used on stone, wood, and bone), and finally in learning all the complex skills connected with the mining and smelting of copper ores, with the casting and hammering of metallic copper, and then the alloying of copper with arsenic and tin.[9]

Viewed in this light, metallurgy appears to have constituted an even more radical innovation than pottery. The original inspiration may have come, as with pottery, from observing processes occurring naturally – after a volcanic eruption, or, closer to home, when the clay used for baking pots contained metal ores that melted. The move to controlling these processes, and to experimenting with the smelting and mixing of various rock substances, was enormous. Not surprisingly, therefore, the activities associated with metalworking were, from the very beginning, the occupation of specialists: miners and smiths.

Even more than the baking of earthenware, metallurgy demanded social organization and specialization. Prospectors had to know where to find the metals, how to recognize them and how to extract them. Then, the proper alloys had to be composed in a crucible, poured into previously prepared moulds, and annealed or cooled. The preferred fuel for heating the crucible was charcoal, which had to be produced in advance. Further provisions included a furnace, stone or clay casting moulds, tongs, bellows, a hammer and an anvil. As the British archaeologist C. R. Wickham-Jones observes, even the simplest form of metallurgy could be accomplished only by 'careful organization, timing and skill at each stage'.[10]

In spite of current reservations about a unilinear sequence of technological advances, all the published evidence still seems to support the general conclusion that metallurgy began with the working of lead and copper, followed by the invention of bronze, an alloy of copper with either arsenic or tin, and later still, by the discovery of ways of working iron. Once techniques had been found to make iron into a usable material, both bronze and iron still continued to be used for a time, but, in most instances, iron eventually superseded bronze – mostly, so it seems, because its ores were more abundant than the scattered deposits of copper and tin which were needed for the production of bronze.

Metal could be made to serve a variety of purposes. Before bronze was invented, the available metals – gold, silver and copper – were mainly used for decorative and ceremonial purposes; although copper pins and fishhooks obviously had practical functions as well. Copper cups and plates had the advantage of being more easily manageable and more durable than earthenware. Being scarce and simple to transport, metal objects were often used as gifts and as a means of exchange. The processing of bronze and iron opened the way to a wide range of applications as tools, vessels and weapons.

More than anything else, the manufacture of weapons became an exclusive specialism of metalworkers. As Colin Renfrew notes, 'until daggers were invented, no metal product was so remarkable or original as to be indispensable'. Nor did any previously designed weapon made of any other material have a combined thrust, sharpness and strength equal to the dagger, or its derivatives, the rapier and the sword. 'The new form ... produced a universal military threat which could be answered only by equipping oneself with similar weapons.'[11]

Not only did metallurgy create, for the first time in history, 'a whole range of valuable objects worth hoarding in quantity'[12] but it also supplied the weapons with which these objects might be appropriated. It reinforced the trend, present in most settled agrarian societies, towards accumulation of property; but it also tended to turn this trend in the direction of a highly uneven distribution of accumulated property. The possession of weapons, which had already for a long time tended to be the monopoly of adult and fully initiated men, to the general exclusion of women and children, now came to be monopolized by one specific class of men: the warriors. Increasingly, as the American sociologist Gerhard Lenski notes, 'the energies of this powerful and influential class were . . . turned from the conquest of nature to the conquest of people.'[13]

The monopolization of the means of destruction by specialized warriors, who were able to command military units equipped with bronze and iron weapons, forced peasants and craftspeople into submission. These peasants and craftspeople had no choice. In order to pursue the productive activities on which their social existence hinged they needed warriors – to protect them from other warriors.

It is sometimes suggested that smiths formed an exception. There are legends about 'royal smiths' who started their careers as humble craftsmen but became, thanks to their special abilities, the founders of dynasties. The most famous example is the great Mongol conqueror Genghis Khan, who was said to have originally been a smith, or at least descended from a family of smiths.[14] There may have been a kernel of historical truth in some of these stories in that initially small guild-like clans may have managed to keep both the secrets of their trade and the combined monopolies of weapons' manufacture and military organization in their own hands. However, even if such combinations of the roles of smith and warrior occasionally occurred, none of them proved viable in the long run.

Just as the peasants supplied the rulers with their food, so the smiths supplied them with their weapons. This sentence brings out both the importance of their social function and the weakness of their position. Their mastery over fire and metal was undisputed and much needed; what they lacked was mastery over people. They lacked the means of organization, which were controlled by military leaders.

In their military operations, the warriors relied more heavily on the products of fire technology (or pyrotechnology) than on fire as such. Fire did not play a direct part in combat – except in sieges, when it was sometimes used to undermine the enemy's defences. It was applied most spectacularly, however, when a siege had been completed with conquest. What the British ancient historian O. R. Gurney notes in his account of the Hittites applies to military-agrarian society in general: 'A city conquered by force of arms was the . . . prey of the victorious army and was generally looted and burned to the ground.'[15] In the chapters that follow I will return to this practice of sacking and burning, and to questions of how and why it was carried out.

FIRE USE AND FIRE HAZARDS IN CITIES

Babylon

Among the specialists in advanced agrarian societies were scribes, often attached to temples. They developed the art of writing, to which we owe records about some aspects of urban life in Mesopotamia stretching back more than 5,000 years. In these sources, fire is rarely mentioned. The French philologist Jean Bottéro, who has collected all references to fire in ancient Mesopotamian texts, considers that there are three conditions to be met before we can claim to possess written evidence about any aspect of life in the ancient world. It has to be recorded in the first place; the recording has to have survived; and it has to have been rediscovered. With regard to fire, even the first condition was largely absent: in the cities from which the records originate the use of fire would have been so common and unproblematic that there was seldom a particular reason to refer to it in writing.

Yet Bottéro has succeeded in bringing to light some interesting facts. Apparently, both the older Sumerian and the younger Semitic Akkadian language had two words for fire. One was a common, prosaic word, represented by the pictogram of a fire pot; the other was much more solemn, and was represented by a reed fire. We can perhaps interpret the two pictograms as references to the two original forms of controlled fire that I distinguished in Chapter 2: the cooking fire (the most thoroughly 'domesticated' form of fire) and the fire used for clearing land (and more closely resembling a 'wild' fire). It is also intriguing in this context to reflect on the fact that all Western European languages have at least two etymologically very different words for fire and for burning. Although in present usage the distinctions have become blurred, it is possible that the original meanings may have been associated with either, respectively, fire as a harmless,

precious object (*feu, Feuer, vuur*) or fire as a dangerously active force
of nature (to burn, *brûler, branden*). This, however, is mere speculation.
As far as I know, the problem has not yet been investigated.

Another interesting linguistic fact is the absence in Mesopotamian
texts of any term for 'making fire'. Bottéro suggests several explanations: it might mean either that making fire was a sacred and
therefore secret activity, or that it was too trivial to deserve mention.
A third, and in my view the most plausible, explanation would be
that in any city there were quite a large number of fires burning all
the time, so that people could always take embers from an existing
fire instead of having to make it themselves by the cumbersome
methods of using fire stones or a fire drill.

What were the uses of fire? Bottéro lists, first of all, a number of
domestic purposes: to heat and light one's home and, at special
occasions, to fumigate it and to fill it with the scent of incense; other
domestic functions included heating water for a bath, and, most
important of all, cooking. The houses did not have chimneys; portable
fire pots or braziers were used for heating, and cooking was done
outside.

A second series of uses of fire concerned the many crafts that were
carried out in the cities, often in the service of the temples. Smith,
baker, potter, brick-maker, glass-maker, brewer, cooper, wood-
worker, basket-maker, boat-builder, leather-worker, wheat-dryer,
dates-dryer, oil-maker, perfumer, minter, charcoal-burner, chalk-
burner, cattle-brander, slave-brander – there was hardly any occupation that did not involve fire. In some crafts, such as those of smiths
and bakers, the furnace or oven formed the centre of the workplace;
in others, such as basket-making or boat-building, it was needed at
certain stages of the work process, as, for instance, caulking seams
with pitch.

The many uses of fire in houses and workshops must have created
fire hazards. Remarkably little is said about this in the sources,
however. The oldest-known collection of written laws, the code of
Hammurabi, recorded around 1780 BC and laying down a set of
rules for everyday life in the city of Babylon, does not contain any
ordinances compelling citizens to take special precautions in the use

of fire. It is as if the elementary care needed in handling fire was regarded as a basic skill which everybody would have acquired as a child, and which did not have to be explicitly reinforced by written law.

The lack of references to problems of fire-prevention is not peculiar to Mesopotamia; it seems to be a general characteristic of texts from the ancient Middle East and the Mediterranean regions prior to the rise of Rome as a metropolis. There are probably two explanations: first, a sense of caution was more or less taken for granted, and second, houses were generally low, with fire-resistant outside walls of stone, brick or mudbrick, and were not built too closely together, so, in the event of a fire, the whole city would not be immediately endangered.

Once a blaze had started, there was little that could be done to save the interior timber and the roof of individual houses. It has been said that water, sand and prayer were about equally effective as ways of fighting fire.[16] This is all the more plausible since water was usually available only in small quantities, which had to be brought to the fire in buckets and then thrown on to it. Neighbouring houses were protected by applying wet cloths to the roofs. If the conflagration was too great to be contained in this manner, the only remaining measure was to create a 'firebreak' by tearing down buildings.

That fires did occur can be inferred from the following rule in the code of Hammurabi.

If fire broke out in a seignior's house and a seignior, who went to extinguish [it], cast his eye on the goods of the owner of the house and has appropriated the goods of the owner of the house, that seignior shall be thrown into that fire.[17]

The fact that this is the only reference to fires in the entire code also suggests that the main concern of the city authorities was not to fight the blaze but to counteract the disruption of public order which it might occasion in the form of thefts and fights. Looting was punished directly, without trial. The harshness of the penalty was in line with the general tenor of the code, according to which many offences were punishable by death. (Execution by fire was rare, however; the only other instance in which it was prescribed was for a mother and son who had sexual intercourse after the death of the father.)

The code also reveals, in passing, that it was customary to brand slaves, like cattle. If a brander removed the brand-mark of a slave who was not his own, his hand had to be cut off. Whoever instructed a brander to remove the brand-mark of a slave not his own was to be brought to death.[18] These regulations seem to point to great differences in power, not just between citizens and slaves but between the city authorities and individual citizens as well.

There is no evidence of a fire cult in Mesopotamia, such as developed in neighbouring Persia. The priests used fire in the temples in the same ways that the citizens used it in their homes: for warmth and light, for fragrance, for heating bath water and for cooking. The use of fire in sacrifice was intimately linked with cooking; the meals dedicated to the gods were in fact eaten by the priests.[19] A very special way of employing fire involved the custom of throwing wax figures of enemies (often declared to be 'witches') into the fire, thus burning these persons in effigy. (Possibly, this practice may help to explain the prohibition in several oriental religions of images of gods and people; it may well have been inspired originally by the fear powerful persons had of being burned in effigy.)

Hattusa

One of the golden rules in archaeology states that 'the absence of evidence is no evidence of absence'.[20] The fact that there is no reference to fire-prevention in the known records of ancient Mesopotamia cannot be taken as proof that the problem did not exist. On the contrary, evidence about early urban life from neighbouring regions clearly suggests that the city authorities were indeed keen to prevent fire, especially when public buildings such as temples and palaces were at stake.

An ancient city about which we do have some evidence on this score is Hattusa, the capital of the Hittites from about 1650 to 1200 BC. Here some unequivocal traces of an official policy of fire-prevention have been preserved. They tell us, for example, that it was forbidden to carry wood or torches into a particular fortress or to light a fire within its walls. In another text, the man who supervised the nightwatch was instructed to shout, during his first round, 'Extinguish the fire!' and, during his next round, 'The fire must be checked!' These instructions apparently referred to the institution of a curfew (or *couvre-feu*) of the kind we know from the cities of Western Europe in the Middle Ages: a general injunction not to let open fires burn during the night. This was an external constraint on individual citizens to do things which, though they might have found them unpleasant and unnecessary, were necessary in the interests of their neighbours and the city at large.

Even more interesting for what it reveals about the links between external constraint and self-restraint is the following instruction for temple servants:

Further: Be very cautious with the matter of fire. If there is a festival in the temple, guard the fire carefully. When night falls, quench well with water whatever fire remains on the hearth. But if there is any flame in isolated spots and [also] dry wood, [if] he who is to quench it becomes criminally negligent in the temple – even if only the temple is destroyed, but Hattusa and the property of the king is not destroyed – he who commits the crime will perish together with his descendants. Of those who are in the temple, no one is to be spared; together with their descendants they shall perish! *So for your own good be very careful in the matter of fire.*[21]

We may infer from this text that the general control by the nightwatch was not considered a sufficient guarantee that all fires would indeed be extinguished at night. Therefore special instructions were given to the temple servants. An appeal was made to their sense of personal responsibility – an appeal made in the form of a stern reminder of the gruesome penalties that would await the offender, his colleagues and his entire progeny. The obligation to exercise caution with regard to fire was presented unambiguously as a social duty, supported by social sanctions.

The laws of the Hittites also provided for the settlement of damages incurred if someone had set fire to another person's property. For example: 'If a free man sets a house on fire, he shall rebuild the house. And for everything that perishes in the house, whether man, cattle, or sheep, he shall certainly compensate.'[22] It would seem that the burning down of a house was not regarded as a terrible disaster; if a free man had caused a fire, all that he was required to do was to pay compensation. Punishment was meted out only if a slave set a house on fire; then his nose and ears would be cut off before he was sent back to his master. The master would have to pay damages; if he refused to do so, he would lose his slave. The rather casual tone of these laws regulating personal retribution among citizens differed remarkably from the thunderous penalties with which the authorities threatened temple servants in the event of a fire.

FIRE IN THE COUNTRY

While from early times cities had written regulations regarding fire, for the peasantry living in the country no such rules appear to have been laid down until quite recently. This was, I think, not only because the peasants were illiterate but also because there was less need for such regulations. Peasants living in villages or isolated homesteads did not have as many different uses of fire as were to be found in the towns. The domestic hearth was rather easy to control, and even if a fire did break out, the damage would usually be limited to a single building.

Depending on the physical environment, houses or huts were built of stone, wood or mud, or a combination of these materials. Roofs were mostly covered with thatch or straw. In 1977 an international team of archaeologists in Serbia came across an abandoned wattle-and-daub house which was similar in its construction to the type of dwelling in which peasants used to live 2,000 years ago. In order to find out what traces would be left if such a building was burned, the archaeologists decided to buy the house and to set it on fire. They deliberately allowed the cooking fire to burn out of control and recorded the results.

Within three minutes, the fire had burned through the roof. Within six minutes, the thatched roof was aflame, and the western chamber with its straw mattress had become an unapproachable inferno. Enormous quantities of smoke billowed from the house. No one working in the surrounding fields could have ignored such a fire.[23]

Almost as quickly as it flared up, the fire subsided. After twenty minutes the thatch was almost entirely burned away, and the roof had collapsed. Only minimal damage had been done, however, to the daub-packed walls and structural elements of the house. As the researchers noted, it would have taken relatively little effort to repair the roof, clear the house of burned debris and restore it to a habitable condition.

Allowing for the great variety in local materials, I think we are justified in seeing the house burned in Serbia as what the French historian Fernand Braudel called, with slight exaggeration, 'an ageless document'.[24] And we can, then, infer from the experiment that, for a long time, peasant houses were constructed in such a way that they were not likely to suffer irreparable damage from fire.

This is not to say that the rural world was impervious to fire. On the contrary, from the time that people began to lead sedentary agrarian lives, and became dependent on stocks of food, fodder and seeds, fire inevitably posed a greater threat to their existence than it had ever done before. The archaeologists who burned the house in Serbia did not fill it with stocks of any kind. If, however, a family's winter provisions had been destroyed in a fire in a peasant's hut, that would have constituted a severe loss. The severity of such losses seems to have increased only as rural populations began to be able, with the aid of pottery and metallurgy, to accumulate more wealth. Intensive growth not only made them more dependent on fire as a productive force but also increased their need to be wary of its destructive potential.

5. FIRE IN ANCIENT ISRAEL

SETTING AND SOURCES

The society of ancient Israel has left us an exceptionally rich heritage of written records, documenting many aspects of its history during the first millennium BC. The major source is the collection of writings that has been canonized in the Christian tradition as the Old Testament. It is, of course, a source that has to be used with some caution. This is especially true of the first books, from Genesis to II Samuel, which claim to describe the history of the Jewish people before the founding of the kingdom of Israel by Saul around 1000 BC. Although these books do contain some very ancient fragments, they date, in the form in which they have been handed down to us, from the time of the Babylonian exile, after the capture of Jerusalem by the Assyrians in 586 BC. There is thus a gap of several centuries between the time when the events which the texts report are supposed to have taken place and the time when the reports as we know them were recorded. Moreover, the authors were not primarily interested in historical accuracy but in designing a model of the past of the Jewish people which would bolster that people's sense of religion and community.[1]

It follows that we can expect two kinds of systematic distortion: first, a strong emphasis on religion and on the part which the god worshipped in this religion was supposed to have played in human history; second, a tendency to exaggerate the unity of 'the Jewish people' as a people with a common patrilineal descent and a common history and culture. Given these biases it is actually astonishing how plausible in many respects is the picture that emerges, and how closely it matches current views of the growth of a settled way of life and of incipient state formation. While there is no historical evidence whatsoever for the stories about the patriarchs, Abraham, Isaac and Jacob, there is nothing our common sense would cause us to doubt in the account given in Genesis of the way they lived, as a semi-nomadic group in the process of changing to a settled way of life. The stay in Egypt and the subsequent exodus led by Moses are also unproven; but again, if we disregard the miracles which adorn the narrative, it is perfectly possible that people living in Palestine, pressed by famine, migrated to the richer land of Egypt, and that a number of their descendants later returned to Palestine. At first they would have formed rather loose tribal federations, and there would have been many outbreaks of violent conflict (as described in Judges), both among the tribes themselves and with neighbouring groups. Around 1000 BC some strong leaders, taking advantage of a temporary weakening of the surrounding empires, succeeded in forging the tribal federation into a kingdom.

The history of the kingdom is told in I and II Kings. Although these too are not entirely free from distortions, they give sufficient historical detail for a reconstruction in terms of modern chronology. Within one century the kingdom fell apart into a northern and a southern half. The north, which continued to be named Israel, was conquered by the Assyrians in 722 BC. In 586 BC the southern part, named Judah, with Jerusalem as its capital, suffered a similar fate under the successors to the Assyrians, the Neo-Babylonians. The books in the first part of what is now known as the Old Testament were edited in the subsequent Babylonian exile.

Two dominant themes in these books are religion and war. This is hardly surprising, given the conditions in which both the heroes whose feats were recounted and the scribes who recorded the tales lived. The setting was that of a military-agrarian society in the Middle East, dominated by large empires which had their centres of power in Mesopotamia and Egypt and whose spheres of influence reached well into the frontier area of what is now called Palestine. The accounts in Joshua, Judges, and I and II Samuel can be read as the history of a number of closely intertwined competing groups trying to gain control over a sizeable part of this territory. After a brief period of military success, the political power of the Israelites waned; even so, they maintained a cultural unity centred on a common religion. At the time of the exile in Babylon, the days when these tribes could celebrate triumphs on the battlefield were over. The feats of bygone days were now recorded with all the more devotion by a literate élite of priests who, though themselves no longer able to wield military power, shared many of the values of the military-agrarian regime in which they lived. They portrayed the past of their own people as a history of conquest and sovereign-state formation, sadly followed by decline and subjugation. The books of the prophets, which were written before, during and also after the return from exile, likewise reflected the responses to military decline, against which the prophets called for a revival of religious discipline and solidarity.

The words fire and burning occur frequently in all these writings, almost always in the context of either religion or war. The authors show little interest in the control of fire as such. Thus, in the scene describing how Abraham prepared himself for the sacrifice of his son Isaac, it is said that he brought along a knife, firing wood and fire (Genesis 22: 6). We may infer that people apparently were not used to making a new fire, even at a distant spot. But clearly this was not the message the author wished to convey. He was interested in the human drama – described as a drama between man and God – of a father intending to kill his child but at the last moment being reprieved from having to do so.

Religious sacrifices on a burning altar form a regularly recurring theme in which fire is mentioned in the Scriptures. In a much more ominous way, fire also appears as a sign of divine power through which the Lord manifests himself to his confidants on earth. A third context for fire is in descriptions of the Lord expressing his divine wrath, striking sinners and enemies of Israel. Closely related to this are accounts of military actions resulting in the destruction of fortresses and towns by means of fire. Finally, as no more than a residual category, there are some casual references to the use of fire for such practical purposes as cooking, heating and lighting.

The control of fire was not a major concern, either to the scribes or to the people of Israel in general. Yet, partly because of the very unobtrusive manner in which they usually refer to it, the texts give us revealing glimpses of the ways in which fire was used in ancient Israel. They present us with a social setting which was in some respects unique (and, as I shall argue, cultivated to be so) and in others reflected features that were generally typical of military-agrarian society in the Middle East during the first millennium BC.

FIRE AND SACRIFICE

Most references to fire have to do with sacrifice. Technically the altar fires posed no problem. The texts are concerned with the far more difficult issue of regulating social conduct in front of the altar – of stipulating everybody's rights and duties – and this is what makes the many relevant passages sociologically interesting. From the dry and legalistic Scriptures emerges a picture showing us how people, in the name of their god, dealt with each other in front of the altar fire, and what sacrifices they were required, and prepared, to make. The perennial problem centred round just who was allowed or obliged to offer what to whom; in order to reach a solution it was necessary also to settle where and when the sacrifice was to be made. All these issues were potential sources of conflict, and the texts show that they sometimes led to fierce clashes.

A dramatic highlight is the story of how Abraham went out to build a fire altar to sacrifice his only son Isaac. He did so, we are told, at the command of the Lord, who wanted to test the strength of his faith. Eventually Abraham did not sacrifice his son because, at the very last moment, he received a message that the Lord would be satisfied with a ram. As a reward for his faith Abraham was promised numerous and blessed descendants.

According to the text, that reward was given to Abraham for his god-fearing conduct as expressed in his willingness to offer, at the command of the Lord, his dearest possession, his son. From a secular point of view, there seems to be a contradiction between the original command and the reward, as the only way for Abraham to have had progeny would have been *not* to sacrifice his son – only he who spares his sons, and resists any possible temptation to kill them, can count on male offspring. Perhaps the story of Abraham and Isaac may be read as a reminder to every father of his duty never to give in to a fatal impulse to kill his son, even though he may be physically capable of doing so, and even if there is no state authority to prevent or to punish him.[2]

There is evidence suggesting that the practice of sacrificing sons, especially the firstborn, did indeed occur. A stricture in Exodus says: 'The firstborn of thy sons shalt thou give unto me' (22: 29).[3] Apart from one other rather ambiguous statement in Exodus (13: 2), this command appears to stand alone in the corpus of prescriptions which, like archaeological layers, represent the sediment of different eras. All later admonitions clearly condemn human sacrifices. Other nations are spoken of with contempt for indulging in such practices, and the Israelites are strictly forbidden to copy them, as they are forbidden to take part in any form of the idol worship of surrounding peoples. Thus Deuteronomy warns the Israelites not to assume the customs of the original inhabitants of the land of Canaan: 'Thou shalt not do so unto the Lord thy God: for every abomination to the Lord, which he hateth, have they done unto their gods; for even their sons and their daughters they have burnt in the fire to their gods' (12: 31).

The service of Molech is especially denounced. The laws in Leviticus emphatically state: 'And thou shalt not let any of thy seed pass through the fire to Molech, neither shalt thou profane the name of thy God: I am the Lord' (18: 21). They go on to add the sanction: 'Whosoever he be of the children of Israel, or of the strangers that sojourn in Israel, that giveth any of his seed unto Molech; he shall surely be put to death: the people of the land shall stone him with stones' (20: 2). The penalty is severe, in accordance with the time in which the rule was written down – a time when, in the absence of a strong state government, the only agency to carry out collective sanctions was the people's tribunal, acting as judge and executioner simultaneously.

It may no longer be possible to ascertain how much exaggeration there was in the repeated claim of the prophets that the Israelites themselves also committed the atrocity of burning 'their sons with fire for burnt offerings unto Baal' (Jeremiah 19: 5). According to Jeremiah, this happened so frequently that he suggested renaming the place where the crime was committed 'the valley of slaughter' (7: 32). Other prophets reiterated the accusation (see, for example, Ezekiel 16: 20, 20: 26, 23: 37–9). Their allegations find some support in the historical books I and II Kings, where it is said that King Solomon built a Molech altar (I Kings 11: 7), and that each of two of his successors burned his own son as an offering (II Kings 16: 3, 21: 6). Apparently it was only towards the end of the seventh century BC that the pious King Josiah put an end to these practices by demolishing the shrines, so 'that no man might make his son or his daughter to pass through the fire to Molech' (II Kings 23: 10).

The campaign against child sacrifices to Molech and Baal bore some resemblance to campaigns in more recent times that have been labelled 'civilizing offensives'.[4] It was conducted by religious leaders who tried to induce the Israelites to obey a set of rules – known as the laws of Moses – which were purported to be of divine origin and of which they claimed to be the sole authorized interpreters. A considerable number of these rules concerned offerings. Offerings were not to be made lightly and just anywhere; the only proper place for them was a consecrated altar. An offering usually consisted of meat or other food; depending upon the type of ceremony, this food was either burned in its entirety or prepared for a meal which, again depending on the type of ceremony, was consumed by the person making the offering or by the priests in charge of the altar.

The offering rituals as they are described may be viewed, it seems to me, as the specific cultural response to problems with which people in settled agrarian societies were confronted generally. As such, the rituals formed part of a more encompassing 'agrarian regime' – a set of ritually binding solutions for problems arising out of the agrarian way of life. One such problem was what to do with the abundance of food available directly after the harvest or after the birth of young livestock in early spring. After a hard winter, there was always the temptation to wallow in abundance; however, in the long run, groups which gave in to this temptation stood less chance of surviving than groups with a more disciplined attitude. Ritual feasts could bolster a more prudent and pious attitude. During these feasts a small part of the new harvest and the young livestock was immediately consumed, while the larger part was saved, as 'invested capital'. It seems unlikely that right from the start all members of an agrarian community would always have been prepared to act so sensibly of their own accord; more probably, certain influential people would have taken the lead and compelled others to follow their example. Such enlightened leaders may well have been the forerunners of the religious specialists, the priests.[5]

Once priests had become part of the established social order, a new problem arose: how to regulate the relationships between them and the rest of the people. A large part of the legislation ascribed to Moses provides solutions to this problem. It lays down the respective rights and duties of priests and of laymen at all sorts of offerings. The fact that the priests depended on these offerings for their livelihood makes it easier to understand why all the actions pertaining to the various sacrificial ceremonies were defined so minutely. While the rules might appear to refer to individual conduct, they actually spelled out social obligations. And while these obligations were presented as eternal laws, they probably reflected the prevailing balance of power between the religious specialists and the rest of the people at the time they were formulated.

There can be no doubt that at that time the altar fire was sufficiently 'domesticated' to mean that making an offering posed no technical difficulties. The emphasis in the texts was therefore fully on the correct performance of the ritual. The main concerns were, first, that the believers should restrict their devotion to the one and only god, the god of the priests of Israel, and, second, that everyone should receive his due: the god, the priests and those who brought the offerings.

Again and again it was reiterated that the offerings were to be made only at the designated spot and that any deviation would be punished severely. In one of the earliest stories it was told how the sons of Aaron offered 'strange fire' — that is, fire not taken from the permanently burning altar in the tabernacle. They were immediately punished: 'And there went out fire from the Lord, and devoured them, and they died before the Lord' (Leviticus 10: 2).

Other texts suggest that it was left to the believers themselves to execute the penalty: 'Whatsoever man there be of the house of Israel ... that offereth a burnt offering or sacrifice, And bringeth it not unto the door of the tabernacle of the congregation, to offer it unto the Lord; even that man shall be cut off from among his people' (Leviticus 17: 8–9). The issue was indeed important. If the offerings were not brought to one central place, the religious authorities would lose control over them. The prophet Jeremiah described the dire consequences this would have: 'The children gather wood, and the fathers kindle the fire, and the women knead their dough, to make cakes to the queen of heaven, and to pour out drink offerings unto other gods' (Jeremiah 7: 18). He spoke with abhorrence of all the houses in Jerusalem 'upon whose roofs they have burned incense unto all the host of heaven, and have poured out drink offerings unto other gods' (19: 13).

Many pages are filled with stipulations of what was due, at the official burning altar, to each of the three parties concerned: the god, the priests and those who brought the offerings. The books do not agree in every detail, which may indicate some shifts in the balance of power between priests and laypeople. On the whole, as a form of taxation in kind, making offerings seems to have become less important as increasing numbers of people came to live in towns and to use money as a means of exchange.

At the same time, we can see a tendency no longer to judge piety by people's performance of ritual duties but by their general attitudes in life. Several passages bear testimony to this trend. Isaiah was the first to let his god sigh: 'Bring no more vain oblations; incense is an abomination unto me' (Isaiah 1: 13). Later prophets were to reiterate: 'Your burnt offerings are not acceptable, nor your sacrifices sweet unto me' (Jeremiah 6: 20); 'For I desired mercy, and not sacrifice; and the knowledge of God more than burnt offerings' (Hosea 6: 6); 'I hate, I despise your feast days, and I will not smell in your solemn assemblies. Though ye offer me burnt offerings and your meat offerings, I will not accept them: neither will I regard the peace offerings of your fat beasts' (Amos 5: 21–2). What the Lord requires of his followers is·'but to do justly, and to love mercy, and to walk humbly with thy God' (Micah 6: 8).

The tendency towards 'internalization' of religion evinced by these words appears to have been a characteristic not just of the development of Judaism but of all great 'world religions', including Christianity. The American anthropologist Marvin Harris has related this trend to ecological changes, especially to the increasing scarcity of meat, as human populations grew in size.[6] It would not contradict Harris's general line of argument to suggest that, as successive generations of peasants became increasingly more accustomed to living under an agrarian regime, the need to bolster that regime with external ritual constraints grew less pressing. At the same time, as more people came to live in cities, the agrarian aspect of their moral-religious regime (their 'civilization') became less relevant for them. These various trends may help to explain why offerings in kind tended to be replaced by saving and by paying taxes – social activities not involving fire.

FIRE AS A SIGN OF DIVINE POWER

In the story of the Creation in Genesis, fire is not mentioned. This appears to indicate that, in contrast with water, fire did not figure prominently in the world view of the scribes who, living in a hot, arid land, wrote down this story. There are, however, several other passages in which we are told how God has manifested himself to certain elect persons 'in fire'.

One such passage is in Exodus, where it is related that God ordered Moses to lead his people out of Egypt into the land of Canaan. On that occasion, the angel of the Lord is said to have appeared before Moses 'in a flame of fire out of the midst of a bush: and he looked, and, behold, the bush burned with fire, and the bush was not consumed' (3: 2). The proclamation of the ten commandments took place in similar circumstances on Mount Sinai, which 'was altogether on a smoke, because the Lord descended upon it in fire: and the smoke thereof ascended as the smoke of a furnace, and the whole mount quaked greatly' – an awesome happening which filled all those present with great fear (19: 18). Moses was able to strengthen his own authority by remaining unperturbed and assuring his followers: 'Fear not: for God is come to prove you, and that his fear may be before your faces, that ye sin not' (20: 20).

Thus the text attributes to Moses the extraordinary power of being able to communicate with a supreme god who, among other things, controls fire, even in the undomesticated guise of burning bushes and coppices, and thunder and lightning. On an earlier occasion, Moses is said to have driven the Egyptian pharaoh into despair by calling down a series of plagues on the country because the pharaoh refused to let the Israelites leave his land. In the seventh plague, so the story goes, all of Egypt was hit by thunder and hail, 'and the fire ran along upon the ground' (9: 23). We are given to understand, then, that it was within Moses' power to make his god send down such terrible visitations. Only after this god in a tenth visitation had killed all the firstborn in Egypt, both children and livestock, did the pharaoh give in and beg Moses to 'Rise up, and get you forth from among my people, both ye and the children of Israel; and go, serve the Lord, as ye have said' (12: 31).

During the exodus from Egypt the Lord is said to have gone before his people continuously, in a pillar of cloud by day, in a pillar of fire by night (13: 22). The interesting question is not so much what actually happened (for all we have to go by is what the text itself tells us) but what the authors had in mind. They might have thought either of heavenly fire, sent directly by God, or of a fire that was kept burning in the holy tabernacle carried by people at the front of the caravan. In the latter case it would have been the same fire that, according to Leviticus 6: 9–13, was always kept burning in the tabernacle and, later, in the temple.

Other stories in which fire is presented as a sign of divine power show a similar ambiguity. They relate miraculous events that are attributed to direct interventions by God but could also conceivably have been brought about by people. This ambiguity also pervades the longest and most dramatic fire episode in the Old Testament, the duel between Elijah and the prophets of the Canaanite god of fertility, Baal.

The story of Elijah is situated in a much later, and historically already somewhat better documented, era than those of Abraham and Moses. Elijah must have lived in the middle of the ninth century, after the partition of the kingdom in 932 BC. He flourished during the kingship of Ahab (875–854 BC) over the northern realm, which continued to be called Israel. There was in those days a strong tendency, especially in higher circles, to assimilate with the surrounding Canaanite people. Thus Ahab had married the Phoenician princess Jezebel, and he officially allowed the worship of the fertility god Baal.

Elijah first makes his appearance in I Kings as the leader of the resistance against these foreign influences. Like Moses, he is credited with miraculous powers. At one time he prophesied that the Lord would punish Israel's apostasy with a drought that would last for years. When, subsequently, the country indeed suffered such a drought, the king accused Elijah of having caused it. Elijah then asked the king to stage a public contest on Mount Carmel. The story of that contest is told so vividly that I quote it in full, adding my own italics to underline the role played in it by fire.

So Ahab sent unto all the children of Israel, and gathered the prophets together unto mount Carmel.

And Elijah came unto all the people, and said, 'How long halt ye between two opinions? if the Lord be God, follow him: but if Baal, then follow him.' And the people answered him not a word.

Then said Elijah unto the people, 'I, even I only, remain a prophet of the Lord; but Baal's prophets are four hundred and fifty men.

'Let them therefore give us two bullocks; and let them choose one bullock for themselves, and cut it in pieces, and lay it on wood, *and put no fire under* : and I will dress the other bullock, and lay it on wood, *and put no fire under* :

'And call ye on the name of your gods, and I will call on the name of the Lord: *and the God that answereth by fire, let him be God.*' And all the people answered and said, 'It is well spoken.'

And Elijah said unto the prophets of Baal, 'Choose you one bullock for yourselves, and dress it first; for ye are many; and call on the name of your gods, but *put no fire under*.'

And they took the bullock which was given them, and they dressed it, and called on the name of Baal from morning even until noon, saying, 'O Baal, hear us.' But there was no voice, nor any that answered. And they leaped upon the altar which was made.

And it came to pass at noon, that Elijah mocked them, and said, 'Cry aloud: for he is a god; either he is talking, or he is pursuing, or he is in a journey, or peradventure he sleepeth, and must be awaked.'

And they cried aloud, and cut themselves after their manner with knives and lancets, till the blood gushed out upon them.

And it came to pass, when midday was past, and they prophesied until the time of the offering of the evening sacrifice, that there was neither voice, nor any to answer, nor any that regarded.

And Elijah said unto all the people, 'Come near unto me.' And all the people came near unto him. And he repaired the altar of the Lord that was broken down.

And Elijah took twelve stones, according to the number of the tribes of the sons of Jacob, unto whom the word of the Lord came, saying, 'Israel shall be thy name':

And with the stones he built an altar in the name of the Lord: and he made a trench about the altar, as great as would contain two measures of seed.

And he put the wood in order, and cut the bullock in pieces, and laid him on the wood, and said, '*Fill four barrels with water, and pour it on the burnt sacrifice, and on the wood*.'

And he said, 'Do it the second time.' And they did it the second time. And he said, 'Do it the third time.' And they did it the third time.

And the water ran round about the altar; and he filled the trench also with water.

And it came to pass at the time of the offering of the evening sacrifice, that Elijah the prophet came near, and said, 'Lord God of Abraham, Isaac, and of Israel, let it be known this day that thou art God in Israel, and that I am thy servant, and that I have done all these things at thy word.

'Hear me, O Lord, hear me, that this people may know that thou art the Lord God, and that thou hast turned their heart back again.'

Then the fire of the Lord fell, and consumed the burnt sacrifice, and the wood, and the stones, and the dust, *and licked up the water that was in the trench*.

And when all the people saw it, they fell on their faces: and they said, 'The Lord, he is the God; the Lord, he is the God.'

And Elijah said unto them, 'Take the prophets of Baal; let not one of them escape.' And they took them: and Elijah brought them down to the brook Kishon, and slew them there. (I Kings 18: 20–40)

There are several ways of interpreting this story. One is to take it literally, and to accept that everything happened exactly as recorded. The second is to reject every claim to the effect that anything of the kind occurred at all. The third, and to me the most promising, interpretation is to believe that the story may contain a kernel of truth: a man named Elijah was in possession of an easily combustible liquid, the secret of which was known only to a few initiates. Confronted by rivalling pretenders to religious leadership, Elijah had recourse to this mysterious concoction. Maybe he put a small piece of obsidian on the altar that could operate as a burning glass under the midday sun, heating the inflammable liquid to the point of ignition. Thus he would have used his firm control over fire in order to regain some of his shaken control over people.

Elijah's feat was less spectacular than that attributed to Moses, who allegedly persuaded his Lord to allow a terrible storm of thunder and lightning to come down over Egypt. It is precisely the more modest proportions of Elijah's miracle, however, that make it amenable to a 'natural' explanation. Conceivably, he was a virtuoso with fire, having acquired esoteric knowledge about it. Even if the priests of Israel condemned magic, this need not mean that religious specialists never used it.[7] A later text, written around 100 BC and included in the Apocrypha, in Maccabees, describes how Nehemiah accomplished a similar feat, making fire out of water. The account of that event offers circumstantial support for the idea that the liquid Elijah had poured over his offering was a petrol compound, the so-called neftar or naphtha. I will return to this episode further on in this chapter.

FIRE AS A SIGN OF DIVINE ANGER

In the stories of how the Israelites were guided to the promised land
in a pillar of fire and smoke, the Lord was using fire beneficially for
them. We also hear of many episodes, however, when he brought
harm with it – not only upon their enemies but also upon the Israelites
themselves, when they had aroused his easily inflammable anger:
'For the Lord thy God is a consuming fire, even a jealous God'
(Deuteronomy 4: 24).

Thus he did not shrink back from sending a rain of 'brimstone and
fire' (Genesis 19: 24) upon Sodom and Gomorrah, to punish the
inhabitants for their depravity. The towns were destroyed for ever;
all over the country the smoke was seen to arise from the earth 'as
the smoke of a furnace' (19: 28). The sons of Aaron, who used
'strange fire' for their offerings instead of fire taken from the perennial
flame in the tabernacle, met with a similar fate: 'And there went out
fire from the Lord, and devoured them, and they died before the
Lord' (Leviticus 10: 2). When, during the exodus from Egypt, the
people complained, the anger of the Lord was kindled, 'and the fire
of the Lord burnt among them' until Moses interfered with prayer
and the fire abated (Numbers 11: 1–3).

In the books of the prophets, relating to the later era of the
kingdoms of Israel and Judah, the stories are not so much concerned
with instances of divine punishment in the past as with menaces for
the future. Thus Jeremiah threatens in the name of his God that, if
the people of Jerusalem do not keep the Sabbath day holy, the Lord
will 'kindle a fire in the gates thereof, and it shall devour the palaces
of Jerusalem, and it shall not be quenched' (Jeremiah 17: 27). Other
prophets repeated the menace literally: 'because they have despised
the law of the Lord ... I will send a fire upon Judah, and it shall
devour the palaces of Jerusalem' (Amos 2: 4–5).

Compared with the monotonously reiterated warnings of such
prophets as Amos and Hosea, the book of Isaiah is remarkable for its
variety of graphic pronouncements about fire: 'And the people shall
be as the burnings of lime: as thorns cut up shall they be burned in

the fire' (Isaiah 33: 12). 'Who among us shall dwell with the devouring fire? who among us shall dwell with everlasting burnings?' (33: 14). 'And the streams thereof shall be turned into pitch, and the dust thereof into brimstone, and the land thereof shall become burning pitch. It shall not be quenched night nor day; the smoke thereof shall go up for ever: from generation to generation it shall lie waste; none shall pass through it for ever and ever' (34: 9–10).

The grimness of these prospects of devastation by fire should not obscure the fact that they form only a small part of the total texts. The catastrophes that the Israelites feared most of all were drought, which would inevitably lead to famine, and war. They were inclined to associate the idea of fire as a penalty from the Lord almost automatically with military destruction – even if the associations were not very realistic, as in Isaiah's vision of the destruction of Assyria: 'And the light of Israel shall be for a fire, and his Holy One for a flame: and it shall burn and devour his thorns and his briers in one day' (10: 17).

A military image is clearly implied in the verses towards the end of Isaiah: 'For, behold, the Lord will come with fire, and with his chariots like a whirlwind, to render his anger with fury, and his rebuke with flames of fire. For by fire and by his sword will the Lord plead with all flesh: and the slain of the Lord shall be many' (66: 15– 16). The final sentences of the book, describing the fate awaiting all apostates, may sound less warrior-like but are no less gruesome: 'for their worm shall not die, neither shall their fire be quenched; and they shall be an abhorring unto all flesh' (66: 24). According to Genesis, the towns of Sodom and Gomorrah were swept away from the face of the earth at one stroke; what Isaiah announced sounds like never-ending torture, like eternal damnation.

Among his successors, Ezekiel came nearest to similarly sweeping pronouncements. In an eloquent passage of several verses he described how the Lord would gather the people of Israel like silver, brass, iron, lead and tin, and then throw them all into the melting furnace: 'Yea, I will gather you, and blow upon you in the fire of my wrath, and ye shall be melted in the midst thereof. As silver is melted in the midst of the furnace, so shall ye be melted in the midst thereof; and

ye shall know that I the Lord have poured out my fury upon you'
(Ezekiel 22: 21–2). It is a frightening vision, which seems to anticipate
later Christian images of hell and purgatory. The fire it alludes to is
caused neither by lightning, nor by volcano nor by acts of war but,
instead, is a melting furnace – a product and instrument of human
industry. Perhaps kiln-type ovens were sometimes used to execute
death penalties. This might, then, be the historical background to the
miraculous story in the book of Daniel (3: 1–30) about the three
pious Israelites who were thrown into a 'burning fiery furnace'
because they refused to worship a golden idol set up by King Nebuchadnezzar, and who came out unscathed – to the great amazement
of Nebuchadnezzar, who reportedly exclaimed: 'There is no other
God that can deliver after this sort.' It is also possible, however, that
Ezekiel had never actually witnessed an execution in a furnace. The
image he evoked, of large crowds of people thrown together like
metal in a furnace, may also have been inspired by the experience of
living in a city and watching a blaze gutting a densely populated
quarter.

FIRE IN WAR

In the Old Testament the themes of religion and war are closely
interwoven. The conquest of the promised land under the leadership
of Moses and Joshua is described in a mixture of religious and military
terms, as a holy war not only waged in the name of the Lord but also
at times fought with his active support (see, for example, Deuteronomy 9: 3); the Lord acted truly as 'a god of war'.[8]

The social setting of the conquest story is reminiscent of the 'primal
contest' model sketched by Norbert Elias in *What Is Sociology?*[9] That
model describes two tribes that have inadvertently come to be in
each other's way and have become entangled in a life and death
struggle. Previously they were not even aware of each other's existence; now they are interdependent, their lives have become fixated
upon one single goal – to get the others out of their way, to make
them vanish. This is also the mentality described in Deuteronomy:

the Israelites find themselves compelled to exterminate the peoples whose land they are invading, and they legitimize this compulsion as the will of their Lord. The 'holy war' is justified by the idea that the other peoples worship strange gods, and are therefore depraved in the eyes of the god of Israel (Deuteronomy 9: 5).

Deuteronomy was compiled at the time of the Babylonian exile, many centuries after the events which it purports to record. Today it is a matter for debate among archaeologists, historians and theologians whether these events ever took place at all, and whether there is any empirical basis for the idea of 'the people of Israel' having entered into 'the promised land'.[10] Maybe there never was a 'conquest', but only a process of settlement of semi-nomads, followed, first by tribe and state formation, and then by incorporation into larger empires, such as Assyria and Egypt. The story of the original 'holy war' would then be an invention by priestly scribes, eager to establish what, in the military-agrarian society of their time, seemed the most respectable legitimation for a nation: conquest and victory under divine command.

The story of the conquest remains of interest, even if it is wholly fictional, for what it tells us about the mentality of the scribes who wrote it down. For them, it seemed self-evident that, after the Israelites had defeated the Amorites, they burned their cities (Numbers 21: 28), and that they did likewise to other peoples (31: 10). Jericho suffered the same fate: 'They burnt the city with fire, and all that was therein: only the silver, and the gold, and the vessels of brass and of iron, they put into the treasury of the house of the Lord' (Joshua 6: 24).

Apparently not all booty was delivered to the common treasury. One man, Achan, had taken a mantle, 200 shekels of silver and a bar of gold for himself, and had hidden these spoils in the earth under his tent. This offence against military discipline was duly detected and punished: the thief was 'burnt with fire' – an exceptional penalty (7: 15).

Having destroyed Jericho, the Israelites received orders, according to Holy Scripture, to do the same to the town of Ai, to take it and set it on fire (8: 8). This they did, making Ai 'an heap for ever, even a desolation unto this day' (8: 28). The last city that Joshua and his men conquered was Hazor: 'They smote all the souls that were therein with the edge of the sword, utterly destroying them: there was not any left to breathe: and he burnt Hazor with fire' (11: 11).

Judges presents an account of the further military struggles of the tribes of Israel before the formation of a unitary kingdom. These struggles were many and violent, as the closing words of the book suggest: 'In those days there was no king in Israel: every man did that which was right in his own eyes' (21: 25). Apart from some seemingly routine references to towns that were conquered and burned down, fire is mentioned only in passing, as in the story of Samson, who caught 300 foxes and drove them in pairs with burning torches tied to their tails into the wheat fields and olive orchards of the Philistines (15: 4–6). Contrary to the claim of at least one modern author, this is by no means the oldest evidence of man-made fire.[11] What makes the story more than just a curiosity, however, is that it suggests an – admittedly not very practical – solution to a problem that may often have occupied people at war: how to set fire to a field under unfavourable, but highly common, circumstances: when the wheat had not yet ripened and there was no wind.[12]

Indirectly, fire played a part in warfare since it was used in making weapons. The story of the struggle with the Philistines is situated in the Iron Age. But it is suggested that, initially, the Israelites did not have iron, while their adversaries did: 'Now there was no smith found throughout all the land of Israel: for the Philistines said: Lest the Hebrews make them swords or spears' (I Samuel 13: 19–22; see also Judges 1: 19).

The idea of the Philistines having monopolized the production of iron weapons is intriguing and makes us wonder how they might have been able to maintain that monopoly. It is far more likely, however, that the person who recorded the story exaggerated the inequalities in military equipment between the Israelites and the Philistines in order to make the victory of the Israelites look all the more impressive and miraculous. And, indeed, the battle tilted the balance of power in the region. Not long afterwards, the united kingdom of israel was founded, which was soon to enter, under David and Solomon, into the period of its greatest military splendour and wealth. Its riches were derived to a large extent from control over iron mines and smelting furnaces, which provided the Israelites with ample supplies of iron for agriculture, trade and war.

The chroniclers showed little interest in the technique of melting and forging – in this respect they reflected an attitude that was highly typical of scribes as literary specialists. They have given us, however, an excellent account of the political intrigues after Solomon's death, and of the subsequent decline of the monarchy, caused by internal divisions and interference from the larger and more powerful kingdoms of Egypt and Assyria. In 586 BC the Assyrians besieged Jerusalem and took it by force. A few days after the conquest, they burned down the temple, the palace and all the great houses, and broke down the city walls (II Kings 25: 9–10). The leading citizens were carried away into captivity to Babylon; among them were 'all the mighty men of valour ... and craftsmen and smiths' (II Kings 24: 14–16).

In 445 BC Nehemiah, living in exile in Babylon as a high palace dignitary, received word that the city of Jerusalem was still dismantled, its gates 'consumed with fire' (Nehemiah 2: 3). The Persian king, who now ruled over Babylon, gave him permission to return to Jerusalem, and to have its walls restored. One of the things he did, according to Maccabees, was recover the sacred temple fire.

When our forefathers were being carried off to Persia, the pious priests of that time secretly took some fire from the altar and hid it in a pit which was like a dry well and shut it up securely so that the place remained unknown to all.

Many years went by, and then, in God's own time, Nehemiah received his commission from the king of Persia and sent the descendants of the priests who had hidden the fire to recover it. When they reported that they had found no fire but a viscous liquid,

Nehemiah ordered them to draw it up and bring it to him ...

After the sacrificial offerings had been placed upon the altar, Nehemiah ordered the priests to sprinkle the liquid over the firewood and over the offerings laid upon it.

When that had been done, after a while the sun, which had been covered by clouds, began to shine, and a great fire blazed up, to the astonishment of all ...

Nehemiah and his followers called the liquid 'nephtar', which means 'purification', but it is commonly called 'nephtai'. (II Maccabees 1: 19–36)[13]

According to the account in Maccabees, Nehemiah's 'recovery' of fire out of a 'viscous liquid' became the occasion for yearly celebrations and praises to God. It would seem that in a society in which fire was generally available and used for a variety of purposes, its ceremonial functions were still held in high regard. Nehemiah's mysterious rekindling of the temple fire clearly added to his prestige.

However, in comparison with the story of Elijah and the prophets of Baal, the report about Nehemiah's feat is remarkably 'secular'. No mention is made of any active interference on the part of the Lord; the main hero in the story is Nehemiah himself, who is supposed to have known by his own authority that the mysterious liquid would ignite spontaneously.

By the time Nehemiah went back to Jerusalem, Israel had long ceased to be a military power. Its people could only dream of a past in which their forefathers had reduced cities to ashes in holy wars. The altar fire in the temple, however, continued to burn – until, in AD 70, soldiers of the Roman occupation forces set fire to the entire temple.

FIRE IN EVERYDAY LIFE

The use of fire in everyday life, irrespective of any explicitly religious or military purposes, is mentioned only occasionally and in passing. When Isaiah needs a strong metaphor for the breaking of faith with the Lord, he makes the comparison with a potter's vessel that is broken into pieces; 'so that there shall not be found in the bursting of it a sherd to take fire from the hearth, or to take water withal out of the pit' (30: 14). It is on the basis of such casual references that we can conclude that people were no longer used to making fire; if they needed fire, they would take a brand from a fire that was already burning.

It seems likely that, by the time the Israelites became settled (and thereby, conceivably, became 'Israelites'), most of the land of Palestine – and certainly the plains – had already been deforested. Not surprisingly, therefore, we read nothing about fire being used for hunting or about slash and burn. There are a few references to bush fires, and to recompense in the event of someone's setting fire to another's stacked or standing grain (Exodus 22: 6), but they are rare.

The same is true of the domestic uses of fire. There are no admonitions to be cautious with fire, nor are any praises ever sung of the hearth as a focus of comfort and sociability. The conditions of the climate and soil were such that, at this stage of socio-cultural development, water tended to be valued much more highly than fire. The only time fire is mentioned as a means to dispel the cold is in a scene that is far from evoking cosiness: King Jehoiakim, seated in his winter house, with a fire burning in the brazier before him, has given orders that a scroll be read to him on which Jeremiah has written down a message from the Lord; each time three or four columns had been read, the king would cut the portion off 'with the penknife, and cast it into the fire that was on the hearth, until all the roll was consumed in the fire that was on the hearth' (Jeremiah 36: 22–3).

The supply of fuel, one of the steady burdens of settled agrarian life, also receives only scanty mention. Cutting firewood, like carrying water, was considered disagreeable menial work, preferably left to slaves and servants. It is said that Joshua mercifully spared the inhabitants of certain towns so that they and their descendants could serve as 'hewers of wood and drawers of water' (Joshua 9: 27). Although wood and charcoal continued to be the most desirable fuels, other materials were increasingly used as well, including cow's dung and human dung (Ezekiel 4: 15), vine branches (Ezekiel 15: 4–6) and broom (Psalms 120: 4).

As in every society, an important function of fire was the destruction of unwanted and 'unclean' matter. From early on, this had to be done 'without the camp' (Leviticus 8: 17). For the people of Jerusalem, the site for burning rubbish was the Kidron valley; there, during periods of tightening of the religious regime, the images of strange cults were burned (I Kings 15: 3; II Kings 23: 4–6). Later, the same function was also given to the Gehenna valley, where, in earlier days, parents had burned their children for Molech. Later still, the name Gehenna became associated with the notion of a site of ever-burning fire to which sinners would be damned – that notion, however, belonged to the world of the New Testament.

In the older books, a picture emerges of a society with a slowly evolving agrarian regime which seems to have absorbed the more ancient fire regime. The relationships to nature remained precarious, but fire did not rank high among the most feared dangers, either in the countryside or in the cities. Drought, mildew, locust – those were the sorts of disaster that could occasion famine, and against which divine aid was sought (see, for example, II Chronicles 6: 28). In so far as people feared fire, it was mainly fire as a result of acts of war; it was their enemies they feared most, rather than the fire.

The people to whom the texts refer were living in a settled agrarian society. They were either, like Abraham, in the process of settling to a sedentary life, or they had already settled. They rarely needed fire to ward off wild animals; the most frequent use they found for it with regard to animals was branding their sheep, and that was done in order to regulate the property relations among people, not to increase their power over the herds.

Only in specialist occupations did fire continue to be a central element. Potters, smiths, charcoal-makers, bakers and other crafts-people used it to process raw materials into socially valuable products. But their technical skills did not interest the authors of the books of the Old Testament.

For them, the main themes were, again, religion and war. During times of war, the vulnerability of settled agrarian society became painfully evident. Unwalled villages and homesteads were defenceless against military bands. Towns might risk a siege, but if they had to surrender they were submitted to the treatment that has become known to us as sacking and burning: the male citizens would be killed, the women and children enslaved, the houses looted and set on fire. The final destruction, almost a ritual completion of conquest, seems to have served several functions: allowing the victors to vent their feelings of revenge and power, making return hard for those of the vanquished who had escaped the carnage and setting an example to other towns.

In times of peace, agrarian communities could maintain a high level of productivity only through a regime of hard work and saving. Everybody was exposed to the pressures of this regime, which, certainly when the methods of agriculture began to be intensified, relied heavily upon the authority of priests. The communal agrarian cult, centred around the offering of sacrifices on a burning altar, helped to restrain people from slaughtering their livestock at will and from too eagerly consuming their supplies of wheat and seeds. In the long run, the functions of the cult as a constituent of an agrarian regime were gradually superseded by other functions; but even then, for a long time, the altar fire remained at its centre.

Looked at from the perspective of agrarian society at large, the priests and prophets, as upholders of what they proclaimed to be the true religion of Israel, represented forces of cultural divergence in a world in which strong tendencies towards convergence were at work at the same time. By staunchly opposing assimilation and worship of 'strange gods', and by insisting that no sacrifices should be made on 'strange fire', they countered the tendencies to cultural 'homogenization' and added an important strand to the cultural diversity of the ancient Orient.

6. FIRE IN ANCIENT GREECE AND ROME

===

SETTING AND SOURCES

Like ancient Israel, ancient Greece and Rome were part of a configuration of interdependent military-agrarian societies which extended, by the first century AD, as far as Britain and Japan. Throughout this configuration, similar tendencies were at work, resulting in both increasing social differentiation within and increasing cultural variation among different societies.

We need only compare Greece in the fifth and fourth centuries BC with Israel in the same period to encounter striking differences in society and culture. Although they were not far apart geographically, the Greeks and the Israelites at that time had very little direct contact, and they rarely referred to each other in their writings. Their common neighbours, the Phoenicians or Canaanites, appear in a very different light, and even under different names, in Greek and Israelite literature.[1]

Israel was a land-locked society; in some of the rare episodes in which the sea played a part in its literature, it was not sailed over, but the waters receded or it was walked upon. Greek society, on the other hand, had a strong seaward orientation; both economically and militarily it depended to a considerable extent on transport by ship. The Israelites had the bad fortune to live in a land corridor which lay within reach of conquest and occupation by larger and stronger states, so that they enjoyed only a brief episode of political sovereignty. The Greeks, by virtue of being situated at the periphery of the oriental empires, were able to maintain military organizations of their own for many centuries, until they were incorporated into the Roman Empire – in which they still had a privileged position. In Israel, following military defeat, an élite of religious specialists tried to propagate a sense of national solidarity and pride among the people. Greece and Rome had no professional priesthood to speak of, and were ruled by an aristocracy of landowners specifically trained in military skills and virtues.

The oldest surviving Greek books, the *Iliad* and the *Odyssey*, attributed to Homer, present a vivid portrayal of this warrior nobility at an early stage in its formation. Although the exact time and location are uncertain, scholars generally agree that Homer's epics take us into the so-called Dark Ages, the period of migration and resettlement following the downfall, around 1200 BC, of the Bronze Age society of Mycenae, and preceding the revival of a literate culture in the eighth century BC. Supplemented by archaeological research and interpreted with notions from modern anthropology and sociology, the Homeric poems provide a lively picture of the Greek warrior aristocracy in the early Iron Age. Likewise, for a somewhat later age, Hesiod's didactic poem *Works and Days* informs us about the life of free peasants.[2]

From the eighth century onwards, the eastern Mediterranean region, comprising the islands and coastal areas of the Aegean Sea, was the scene of increasingly rapid growth of population and capital. Towns emerged, and developed into *poleis* or city-states. Out of the rivalry between the expanding city-states grew progressively larger military units, at first dominated by the city-states of Athens and Sparta, then brought under the control of the Macedonian princes Philip II and Alexander the Great, and eventually absorbed into the Roman Empire, which, in the first centuries AD, comprised the entire Mediterranean basin and a sizeable part of Asia Minor (including Palestine or Israel) and of Western and Central Europe.

Initially, Rome had been no more than the capital of a small military-agrarian nation, comparable to Jerusalem at the time of King David or Athens under Pericles. In the long run, however, it emerged as the victor in the competition and elimination contests between the Mediterranean capitals. The final triumph over its major rival, the north African city of Carthage, which was completely defeated and consigned to the flames in 146 BC, was one of the milestones in establishing the hegemony of Rome. The military power of the Roman aristocracy remained virtually unbroken during 1,000 years of expansion and consolidation. Its eventual demise became evident when the Visigoths sacked the city of Rome in AD 410 – a dramatic event which formed the occasion for Saint Augustine to write his treatise on *The City of God*.

Altogether, the era of Graeco-Roman history encompasses approximately 1,500 years, from the days of Homer's hero Odysseus to those of the Christian church father Saint Augustine. The great changes in society over that period are reflected in the nature of the written sources. Odysseus is still a semi-mythical figure, and it is unknown whether he was modelled on any particular historical personage. The life of Saint Augustine, on the other hand, is so well documented that his biographers are able to trace his whereabouts and his career from year to year. Yet, even for the later period, information about such elementary matters as the size of the population of cities and provinces is still vague, and it is particularly difficult to find comprehensive data concerning the use of fire. Whereas a good deal has been written about food supply in the city of Rome, not much is known about the supply of fuel.

This is not to say, of course, that fire was unimportant in the world of ancient Greece and Rome; on the contrary, its uses were widespread and varied. Its significance was recognized in religion and mythology, as well as in the more secular philosophies of nature. The Greek pantheon included Hestia, the goddess of the hearth, and Hephaestus, the god of the fiery furnace of smiths and potters; their Roman counterparts were Vesta and Vulcanus. Several early Greek cosmologists pointed to fire as a major force in the universe; later writers dwelt on its great importance for human civilization. At first, a favourite theme was the story of Prometheus, the legendary hero who stole fire from the gods and gave it to humanity. In time authors, most notably the Roman poet Lucretius, and the architect Vitruvius, wrote more secular accounts of the domestication of fire. In the first century AD, the Elder Pliny concluded an overview of craft and industry in his *Natural History* with the remark that 'There is almost nothing that is not brought to a finished state by means of fire.'[3]

Yet in spite of these unequivocal affirmations of the importance of fire, its various uses and hazards were never made into the subject of a comprehensive empirical inquiry. This condition is also reflected in the secondary literature by modern historians. Whereas classical scholars have written at length on the theme of fire in myth and religion, relatively little attention has been paid to the ways in which people actually used fire.[4] I do not intend to pass by mythology altogether, but I will be mainly concerned with what it tells us about the place of fire in the real social world, not just in the world of the imagination.

Fortunately, the original classical writings and the secondary literature do contain many valuable insights and data. What is lacking is an overall survey (to provide one would require a book in its own right). Going through the available literature is like gathering firewood in a forest: there is much to be found, but it is unordered and of varying value. What I have done is to choose a few topics that may help to show how people used fire and how they coped with the problems it posed.

FIRE IN THE WORLD OF ODYSSEUS: THE MILITARY REGIME

The setting of the oldest Greek literary works, the epic poems the *Iliad* and the *Odyssey*, is that of a rural world, dominated by a minority of warriors who were able to mobilize and command organized bands of men specializing in fighting. The commanders were also the owners of landed estates, and their soldiers were recruited from the peasantry. The most important change that was to affect this military-agrarian structure during the 1,500 years of Graeco-Roman history concerned the degree to which the capacity to command organized violence was centralized. In the Homeric age, petty warlords were still able to maintain a large measure of autonomy. In later centuries, armies became increasingly larger, and although even at the heyday of the Roman Empire central control over the armies never ceased to be arduous, the degree of centralization was incomparably greater than it had been even in the days when the city-states of Greece united against Persia.

In the world of Odysseus, as the ancient historian M. I. Finley has named it, social life was centred on the *oikos* – the large family household that was both the major economic and the major political unit. The *oikos* was self-sufficient to a high degree; it provided its own food, wool, leather, wood and stone. The one most-needed commodity that usually had to be imported from outside was metal, which had become indispensable for the manufacture of agricultural tools, jewellery and, especially, weapons. As long as money for carrying out transactions was lacking, the only way of obtaining metals was through exchange or seizure. In the absence of money, exchange took place by bartering goods and services; the only defence against seizure lay in the fighting prowess the *oikos* itself could muster.[5]

The need for military protection gave an immediately visible social function to the ruling upper stratum of warriors. This may help to explain their prominent place in Homeric society, and in the Homeric poems, the setting of which is the war of Greece against Troy. Even when the poet is describing a restful scene of heroes sitting around a fire and enjoying their meal, each man having piously sacrificed a piece of meat to the gods, the background is always the continuing siege of Troy.

In a meticulous study the French philologist Louis Graz has analysed all the places in the *Iliad* and the *Odyssey* at which the word fire occurs. In many cases it is used as a metaphor, as when a battle is compared to a fire 'which suddenly breaks out and falls on a city to set it alight, and the houses go down in the huge blaze, as the force of the wind blows it roaring' (*Iliad* 17: 738–40).[6] Similarly, the way in which an individual warrior throws himself into the turmoil is compared to a bush fire raging on a mountain slope (*Iliad* 20: 490–94) – an image that seems quite fitting to battles which were being waged with ample display of fiery courage rather than in cool calculation. However, according to Graz, the word fire is given its richest meaning when it serves as the symbol for victory and defeat.[7]

The readers and listeners to the poem knew the ending: Troy was doomed to perish, and to suffer 'all the miseries that come on people when their city is captured – the men are slaughtered, fires raze the city, and other men carry away the children and the deep-girdled women' (*Iliad* 9: 590ff.). The story of the eventual conquest and devastation is not told in the *Iliad* (this was done at length much later, by Virgil in the *Aeneid*), but from the second of the twenty-four 'books' (or chapters) onwards, there are recurrent allusions to the ravaging fire which will set the city ablaze and devour it completely (see, for example, *Iliad* 2: 414ff., 20: 312).

We are also reminded several times of the great danger, barely averted, that the Trojans would set fire to the ships of the Greeks. For the Greeks, this would have been a disaster: their fleet was the basis for their operations, and their only means of escape in the event of defeat. At one time the enemy had almost achieved their aim: one of the ships was already ablaze and only at the very last moment were the Trojans, armed with burning torches, forced back by Patrocles and the fire in the half-burnt ship extinguished (*Iliad* 16: 122–304).

Thus, while fire did not play a part in the actual battles, which were fought mainly as man-to-man duels, it was greatly feared as the means of final destruction after defeat. It was not fire blindly ignited by natural causes that was feared, but fire used on purpose by the enemy to complete their victory.

The fires maintained by one's own troops gave little cause for anxiety. One day at the beginning of the war, when the Trojans were still doing well, they struck a heavy blow against the Greeks; that same night they gathered piles of wood and burned many fires all night long, to prepare a feast with sacrifices and to prevent the Greeks from making a furtive escape in the dark (*Iliad* 8: 508ff.).

Apart from cooking, heating and lighting, fire was also used by both sides for the cremation of heroes who had died in battle. The greater the hero, the higher his funeral pyre. The penultimate chapter of the *Iliad* is devoted almost entirely to the solemn preparations being made for the cremation of the body of Patrocles, Achilles' friend, who had been killed by Hector. In revenge, Achilles slaughtered twelve sons of Trojan noblemen and threw them on the pyre, calling to his dead companion:

There are twelve noble sons of the great-hearted Trojans with you in the fire and it is consuming them all together with your body. But for Hector's devouring, the son of Priam, I shall not give him to the fire, but to the dogs. (23: 182-5)

Cremation is a use of fire which I have not touched upon before, although clearly its origins go much further back than the age of Odysseus. Like burial, it is often assumed to be a sign of human religiosity, manifesting itself as early as the Upper Palaeolithic. I should like to suggest an alternative interpretation, a hypothesis more in line with the general approach I am following in this book. It has been observed that predatory animals and scavengers can acquire a taste for human flesh (it was upon such observations that Bruce Chatwin based his speculations about the *Dinofelis* mentioned in Chapter 2). As recently as 1918, after the great influenza epidemic which killed many millions of people all over the world, a leopard in Rudraprayag, India, took to scavenging human bodies that had been left lying around because there were not enough people to cremate them. While leopards generally learn at an early age either 'to avoid contact with people or to treat them with great circumspection', this particular specimen, from then on, continued to prey upon humans, reportedly killing 126 people between 9 June 1918 and 14 April 1926.[8] In the light of these observations, I think we should not rule out the possibility that burial and cremation were invented by our ancestors as a means of stopping scavenging predators from acquiring a dangerous taste for human flesh.

By the time of the Trojan War, it was common practice for people to dispose of their dead by either burial or cremation. The preference for one method or the other was, as the Greek historian Herodotus already knew, a matter of cultural variation. These variations were likely to be influenced, in turn, by such factors as the conditions of the soil and the abundance or scarcity of firewood. According to the historian of ancient Greek religion Walter Burkert, the change from burial to cremation was 'the most spectacular change since Mycenaean time'; it was followed, from the eighth century BC onwards, by a gradual return to interment. Burkert suggests that these changes can be explained only by 'possible external factors – such as wood scarcity – or simply unpredictable fashion'.[9]

For the Greeks and the Trojans at war, cremation was the only honourable way of performing the last rites. For that reason, Achilles was persuaded by his companions to return Hector's body to his people so that it could be properly incinerated. The Trojans spent nine days bringing in vast quantities of wood for the pyre; on the tenth day, they put Hector's body on top of it and set it on fire (*Iliad* 24: 787ff.). The number of nine days is obviously intended to impress the reader with the magnitude of the pyre, but it may also be taken to indicate how scarce fuel had become after a ten-year siege.

With the cremation of Hector, the *Iliad* ends. The royal son of Troy received the last honour that befitted him. At the same time, the fire also foretold the destruction of his vanquished city.

In the *Odyssey*, describing Odysseus' long journey home from Troy, there are no more horrific visions of devastation by fire. One of the hero's first exploits after leaving Troy was the sacking of Ismarus, the city of the Cicones. He killed the menfolk and divided 'the women and the vast plunder' among his own men so that every one received 'his proper share' (*Odyssey* 9: 40–43). The entire episode, told in a few lines, gives us a glimpse of the military regime exercised by Odysseus – he has no mercy towards his enemies, but gives his own men a fair deal and thus creates loyalty among them. Fire is not mentioned at all. On the few occasions it does appear in the *Odyssey*, this is in a context of conviviality and hospitality. Thus we read how Odysseus arrives in a splendid palace hall, lit by flaming torches (7: 101); he receives a hearty welcome and is invited to take a seat of honour in front of the hearth (7: 167). Later, in a more modest setting, a friendly swineherd slaughters two young piglets for him, roasts the meat and serves it up 'piping hot on the spits' (14: 76). In such scenes, fire has only pleasurable associations.

FIRE IN THE WORLD OF HESIOD: THE AGRARIAN REGIME

Ancient society was, as the American medievalist Lynn Whyte noted, 'agricultural to a degree which we can scarcely grasp'.[10] Most people lived in rural areas. Even in the heyday of Rome, a large part of the urban population was composed of immigrants from the provinces, and the ruling families all derived their wealth from land ownership.

At the dawn of Greek and Roman history, the age of original deforestation and slash and burn was long past. Even the earliest writers, Homer and Hesiod, never referred to it. On the occasions when classical authors did mention a forest fire, it was usually as a metaphor recurring as a literary topos. What they had in mind in using this metaphor was a natural fire, caused by lightning or, as a persistent legend would have it, by the rubbing of tree branches in the wind. Wood had become far too precious just to be burned away.

The oldest surviving treatise on Greek agriculture, Hesiod's *Works and Days*, written towards the end of the eighth century BC, is therefore virtually silent on the subject of fire. It is a didactic poem, presented as an attempt on the part of the author to explain to his spendthrift brother the virtues and benefits of leading a frugal and trustworthy life. In a more general sense, it can be seen as an early book of manners, intended for the emerging class of independent farmers who lived in relative security from military and fiscal domination by warriors and tax collectors. As such, it represents a rather exceptional strand in the civilizing process.

Fire, not surprisingly, was of little interest to Hesiod. Among numerous pieces of practical advice, he reminded his brother of his duty to 'burn the glorious thigh-bones' to the (unnamed) 'immortal gods' (337) – an act of piety which the Greek peasants, unlike their Israelite counterparts, were apparently supposed to perform on their own, without priestly supervision.[11] They also had to treat the hearth fire at home with some respect: 'Do not lie down beside the fire when you have just made love, and show your naked parts' (734–5). Hesiod was at his most cautionary, however, in urging his reader not to join the idle men who, in winter, would gather in the smithy and waste their time in front of the furnace instead of doing useful work at home (493–5).

This, indeed, was the main message of *Works and Days*. The poem was directed at free farmers who had to impose a regime of hard work and thrift upon themselves. Hesiod began by telling his readers that they were not living in a golden or a silver age, but in an age of iron, in which 'men work and grieve unceasingly' (176). Much of the ethic of work and thrift that the sociologist Max Weber was later to label 'the Protestant ethic' was already formulated by Hesiod.[12] We may regard him as the oldest-known ideologue of a free peasantry. In that sense, he advocated a specific set of civilizing constraints that might help free peasants to cope with the problems of their social existence. Of course, he was not the inventor of this particular variant of the agrarian regime, but he was an eloquent spokesman for it.

The typical farm which Hesiod had in mind was worked by a man, his wife, a slave and an ox. While such small family holdings may have been viable for a number of generations in some parts of Greece, in the long run they were superseded by larger estates.[13] Consequently, later farming manuals were no longer directed to independent peasants but to landowners and estate managers. The genre as such continued to exist, and it continued to have a strong 'civilizing' tenor, as in Xenophon's dialogue *Oeconomicus* (the *Estate-Manager*), which combined technical instruction with the moral advice to be prudent and not to give in to gluttony or other 'stupid and costly ambitions' (1, 22). But neither the practical treatises on agronomy nor the bucolic poems about rural life, such as Virgil's *Georgics*, treated fire as an important agency in the productive process.[14] The authors tended rather to see its dangers, especially in the hands of careless or malevolent slaves.

THE AGE OF THE GREAT GREEK WARS

By the beginning of the fifth century BC, the Greek cities on the Asian side of the Aegean Sea had come completely within the sphere of influence of the expanding Persian Empire which, by that time, extended as far as Libya and Macedonia. Herodotus, the great traveller and chronicler of the history of his own age, describes how, in 499 BC, the citizens of the Ionian city of Miletus revolted against the Persian occupants. They were soon joined by forces from other Greek cities, which jointly succeeded in conquering Sardis, the former Lydian capital, now serving as headquarters for the Persians. 'But', Herodotus continues,

they were prevented from sacking the place after its capture by the fact that most houses in Sardis were constructed of reeds, reed-thatch being used even on the few houses which were built of brick. One house was set alight by a soldier, and the flames rapidly spread until the whole town was ablaze. The outlying parts were all burning, so the native Lydians and such Persians as there were, caught in a ring of fire and unable to get clear out of the town, poured into the market-square on either bank of the Pactolus, where they were forced to stand on their defence. (5: 102)[15]

'In the conflagration at Sardis', Herodotus added, 'a temple of Cybele, a goddess worshipped in that part of the world, was destroyed, and the Persians later made this a pretext for their burning of Greek temples.' Apparently, Herodotus considered the burning down of temples an act of sacrilege, a greater atrocity than burning down a city. He suggested that while the Greeks committed this atrocity inadvertently, it was later repeated deliberately by the Persians, in vengeance. In mentioning these Persian reprisals, Herodotus anticipated the great wars of 490 and 480–479 BC, when the Persians sent a force of many tens of thousands of men with the intention of quenching the revolt of the cities on the Ionian coast and subjecting the Greek cities on the European mainland as well. In the first expedition they succeeded in reconquering the Ionian cities, and, according to Herodotus, 'once the towns were in Persian hands, the best-looking boys were chosen for castration and made into eunuchs; the handsomest girls were dragged from their homes and sent to Darius' court, and the towns themselves, temples and all, were burnt to the ground' (6: 32).

In the second war, again, everywhere the Persian army went 'there

was devastation by fire and sword, and towns and temples were burnt'
(8: 32–5). One city after another was reduced to ashes, as the Persians
proceeded towards Athens, where, after a heavy siege, they took the
main temple on the Acropolis, 'stripped it of its treasures, and burnt
everything' (8: 52–3).

In spite of this catastrophe, the Greeks were able eventually to
resist the attack of the Persians, and to drive them out of Europe. For
Herodotus, the encounter between the two great forces was the
occasion to inquire not only into the background and the course of
the conflict but also into what we today would call the national
character of the two contending parties and of other peoples in the
area. He dwelt with relish on the cultural variations which he noticed
during his journeys. Especially in Egypt, he was particularly keen on
finding customs which contrasted sharply with what he found the
typically Greek way of doing things.

Herodotus could not know that after their victory over Persia, the
Greeks were to enter into an exhausting internecine struggle, led by
Athens and Sparta, for hegemony over all Greek city-states. The
events which led almost inevitably to the outbreak of that war, and
the first twenty years of its progress, were described by Thucydides
in *The Peloponnesian War* – still a masterpiece not only of
historiography but of social science in general.

In the opening chapters Thucydides described how Sparta and
Athens drifted towards war, drawn into it by a series of events they
could not control. Once the war was under way, it led to a severe
breakdown of public morality; it had what might be called 'decivilizing' effects.

In the few episodes in which fire was used, this occurred in a
typically fumbling way, with results that were mostly unforeseen and
unintended, and sometimes downright catastrophic. Thus, at the
beginning of the war in 429 BC, the Spartans tried to capture in a
swift raid the relatively small town of Plataea. After attempts with
conventional siege engines had failed, they decided to try fire.

They brought up bundles of wood and dropped them down from the
mound first into the space between it and the wall. Since so many people

were taking a hand in the work, this space was soon filled up, and so they went on to heap up the bundles of wood as far inside the city as they could reach from the top. They then set fire to the wood, using sulphur and pitch to make it burn, *and produced such a conflagration as had never been seen before*, or at any rate greater than any fire produced by human agency; for of course there have been great forest fires on the mountains which have broken out spontaneously through the branches of trees being rubbed together by the wind. (2: 77)[16]

It is interesting that Thucydides called the conflagration bigger than anything ever seen before; this suggests that the Greek cities of his time were not often hit by large blazes. The remark about forest fires being caused by branches rubbing in the wind seems to have been added as an afterthought and was probably based on traditional folklore rather than personal experience, for according to modern foresters this is an unlikely way for a forest fire to begin.[17] What Thucydides wished to note was that the fire created by the Spartans was indeed a huge one, which 'very nearly finished the Plataeans off ... and made a large part of the city quite untenable'. As luck would have it, however, the wind abated and a thunderstorm with a heavy fall of rain put out the fire. The city was saved, and the Spartans saw themselves forced to continue their siege, which was to last until the winter of 428–427 BC. When eventually the defenders had to leave, most of them managed to escape at night with the aid of fire signals which confounded the fire signals of their besiegers (3: 22–4).

In a few other episodes fire was also used, but usually the weather interfered and the desired effects failed to occur. At one time, fire worked out favourably for the Athenians when, having landed on an uninhabited island, they were unable to detect their enemies, who were hidden in the dense woods. Then a soldier, cooking his meal, accidentally set fire to the wood, the wind got up and nearly all of the forest was burnt down, revealing the enemies' position (4: 29–30).

There seems to have been only one occasion on which fire was used deliberately and with success. This was when the Spartans laid siege to the city of Delium. They constructed a big tube of wood and iron, which was connected to a cauldron filled with lighted coals, sulphur and pitch. By means of large bellows, they were able to use this contraption as a flame-thrower and thus to set fire to the fortifications, forcing the defenders to abandon their positions and flee.

Thucydides' detailed description of the flame-thrower suggests that it was a unique weapon. During later wars in the Graeco-Roman world, similar weapons were occasionally used – for example, in Alexander the Great's siege of the Phoenician city of Tyros.[18] However, it seems that fire was used only under very specific circumstances, when more conventional means had failed. The main reason for this was probably the unpredictability of the weather, for rain might render fire ineffective in destroying the fortifications of the enemy, while a storm might suddenly turn the flames towards one's own lines.

The operations of invading armies used to include what the ancient historian Victor D. Hanson has called 'agricultural devastation': the wholesale plundering and destruction of the countryside. Fire was a favourite means for destroying houses and property and, in early summer when the grain was ripe, the fields as well. The burning and plundering were carried out by lightly armed and unarmed troops, which were always vulnerable to sorties by cavalry and hoplite troops. The mountainous Greek landscape also restricted the range and speed of their operations. For these reasons, as Hanson points out, devastation was 'a slow process'; it did not just spring forth from a spontaneous fury, but rather formed part of a well-planned strategy aimed at depriving the enemy of his resources and undermining his morale.[19] Permanent destruction of farmland was seldom if ever achieved; since farming depended mostly on labour, the capital losses brought about by military ravaging could usually be replaced within a few years.

Although armies did their best to devastate the countryside, their final aim was to capture the cities. As the military historian Martin van Creveld observes, 'A country was not really occupied until its fortresses had been reduced.'[20] At the time of the Peloponnesian War, this meant that the besieged cities had to be captured and destroyed. The failed attack on Plataea, and other episodes related by Thucydides, suggest that destroying a city by fire was not always easy, not even after it had been conquered. Often enough, extra fuel was needed, such as dry brushwood, straw or oil, in order to consign every house to the flames.[21]

The military uses of fire did not change much when the Romans established their hegemony over the Mediterranean. According to a popular but dubious legend, during the siege of Syracuse by a Roman fleet in 212 BC, the Greek scientist Archimedes constructed a gigantic mirror to reflect the rays of the sun and set the enemy ships on fire. When, in 146 BC, the Romans definitively conquered Greece, they created a certain amount of goodwill by sparing Athens. On other occasions, however, they did not hesitate to destroy enemy cities by fire. The different social profiles of those cities are reflected in the varying emphases put by historians on the destruction of the royal palace in Persepolis, the houses in Carthage and the temple in Jerusalem.

FIRE USE AND SOCIAL STRATIFICATION

Although ancient Greece and Rome were military-agrarian societies, the use of fire was not, of course, restricted to agriculture and war. Indeed, its manifold functions – domestic, industrial, ceremonial – mirrored the far-reaching differentiation of society.

Whatever changes occurred in the control of fire use were mostly the outcome of changing social relations. Technical innovations were of secondary importance, and ideas that might have led to radically new applications failed to have any 'multiplier effect'. Thus, when Hero of Alexandria, in the first century AD, constructed a device in which the expanding hot air produced by an altar fire in a temple was used to open the temple doors, this was seen as no more than an ingenious gimmick; it was certainly not hailed as the discovery of a new principle of converting thermal energy into dynamic energy. The historian of ancient technology K. D. White goes so far as to conclude: 'The control of fire ... remained, for an immense period, completely innocent of technical innovation.'[22] This is not to deny a number of new applications of fire, but these failed to lead to a major

breakthrough in pyrotechnology, comparable to the first development of pottery and metallurgy.

In order for metal to be available for processing by fire, its ores had
to be mined and smelted first. The mining operations involved a high
degree of specialization and social differentiation. According to the
historian Alison Burford, these operations exhibited 'the brutality
prevalent at certain levels of ancient society'. Large gangs of unskilled
workers – slaves, prisoners of war and convicts – were chained
together by shackles, and were 'driven by incessant cruelty to shift
material from the mine-gallery to the surface, or to smash up ore for
smelting'.[23] Of course, mining could not depend on unskilled forced
labour alone; as Burford notes, the operations also required expert
knowledge and organization, represented by a minority of skilled
foremen and overseers. There can be no doubt, however, that the
great mass of miners lived a wretched life at the bottom of the social
pyramid in a highly stratified society.

Smiths and potters were better off than miners. Significantly, the
smiths had their own god, called Hephaestus in Greek and Vulcanus
in Latin. Hephaestus differed from the other Olympians in that he was
disfigured: he limped. Most interpreters agree that this disfigurement
reveals something about the ambiguous status of the smiths. According to some, the maiming of Hephaestus reflected conditions in
archaic times, when the services of the smiths as the manufacturers
of swords were so highly valued that their masters deliberately lamed
them in order to prevent their running away to another lord. An
alternative interpretation is to regard the god's physical appearance
as corresponding to the contemporary stereotype of smiths as men
who manifested physically the dangerous and deforming nature of
the work.[24]

Working with fire, both potters and smiths always ran risks. They
themselves could suffer injuries; and their products could come out
damaged. They had no instruments for gauging the temperature
inside their ovens, nor even for exact time measurement. They had
to rely on their experience and, if necessary, they had to open the
oven in order to see and feel how the fire was going. This could
result in serious burns or even blinding, and it remained uncertain

how the objects to be produced would come out of the firing process. 'If the fire was too hot, or not hot enough, dire things could happen to both the pots in the kiln and the metal in the crucible, or to the object being annealed and worked with tongs and hammer.'[25] As Burford suggests, the uncertainty of their craft made potters and smiths inclined to 'clutter up' their workshops with magical charms. She sees it also as a reason why these craftsmen refrained from experimenting, and stuck to the old, familiar ways of doing things.[26]

Because of their low social status, most potters and smiths have remained anonymous. Yet at times some of them had a chance to distinguish themselves. Hesiod, in the opening verses of *Works and Days*, had already sung the praise of peaceful competition among potters as an example of 'good strife', leading to prosperity, in contrast to the 'cruel strife' of war. In the fifth and fourth centuries BC, the city of Athens regularly staged contests for potters as a part of larger festivals in which an individual such as Bacchius gained fame for his craftsmanship 'in the art of compounding earth, water and fire'.[27]

The uses of fire for domestic purposes varied enormously according to social rank and wealth. In his classic study of *The Ancient City*, first published in 1864, the French historian, Fustel de Coulanges, wrote:

It was a sacred obligation for the master of every house to keep the fire up night and day. Woe to the house where it was extinguished. Every evening they covered the coals with ashes to prevent them from being entirely consumed. In the morning the first care was to revive this fire with a few twigs. The fire ceased to glow upon the altar only when the entire family had perished; an extinguished hearth, an extinguished family, were synonymous expressions among the ancients.[28]

In assessing this statement, which describes customs similar to those still being practised by Brahmins in India today, we have to realize that it pertains only to those who were indeed 'master of a house'. Slaves were excluded, and so was the mass of the urban poor, who lived in cramped apartments in one of the large tenement houses, or *insulae*, of Rome.

In these tenements, the possibilities for fire use were very limited. The apartments contained neither fireplace, oven nor chimney. The tenants had to cook their meals on a portable stove or brazier. Sometimes the rental contract forbade the burning of any stove or brazier inside the flat, because of the fire hazard, and in those cases the tenants had to buy their meals elsewhere.

Only a wealthy minority could afford to live in a private house, a *domus*. As a rule, a *domus* would be far more spacious and better equipped than a flat in an *insula*. There would be an altar fire, as indicated by Fustel de Coulanges, and in addition there would be an oven for baking, stoves for cooking, braziers for heating and sometimes a system of floor heating through pipes – the hypocaust, an ingenious invention that was from the first century AD onwards also used in the public baths.

As the French historian Jérome Carcopino observed in his book on daily life in ancient Rome, great contrasts in light prevailed between day and night. By twentieth-century standards, even the houses of the wealthy were lit poorly by fuming torches and oil lamps. Outside, in the streets of Imperial Rome, there was no lighting at all.

When there was no moon its streets were plunged in impenetrable darkness. No oil lamps lighted them, no candles were affixed to the walls; no lanterns were hung over the lintel of the doors, save on festive occasions when Rome was resplendent with exceptional illuminations ... In normal times night fell over the city like the shadow of a great danger ... Everyone fled to his home, shut himself in, and barricaded the entrance. The shops fell silent, safety chains were drawn across behind the leaves of the doors; the shutters of the flats were closed and the pots of flowers withdrawn from the windows they had adorned.[29]

The rich ventured out at night accompanied only by slaves who carried torches to light and protect them on their way through the mazes of 'unnamed, unnumbered, unlit streets'.[30] If such armed escort was one of their privileges, another was that the houses in which they lived were, apart from all further advantages of space and comfort, less liable to be gutted by fire. The larger *domi* stood in their own grounds, surrounded by walls; the walls helped to keep intruders away, and also served as protection against fire.

FIRES AND FIRE-FIGHTING IN THE ROMAN WORLD

With the growth of cities, having and handling fire became both easier and more risky. There were so many fires that one need not fear that they would all go out; but given the high concentration of people and property, they also constituted a permanent danger, requiring precautions at all times. In the cities of ancient Greece and Rome, as in Babylon and Hattusa in a previous era, these precautions would take the form of a combination of formally issued and officially sanctioned rules – informal 'external' controls that people exercised over each other and 'internal' habits of caution with fire that they had picked up as children and kept almost automatically, as a matter of course.

In all likelihood, Greek city-states had some formal legislation regarding fire prevention. It does not seem to have been considered very important, however. None of the rules seem to have survived, and no reference was made to them by the great political theorists Plato and Aristotle in their treatises on the organization of the state. The secondary literature on ancient Greece also passes over this aspect of city life. Consequently, almost everything we know about fire-prevention and fires in Greece has to do with war, except for a few anecdotes relating to temple fires. Thus Thucydides tells of a priestess who put a lighted torch near the garlands in a temple and then fell asleep, with the result that the temple burned down; the priestess fled that very same night from fear of punishment (4: 133). A more famous story is about Herostratus, who was said to have set fire to the famous temple of Artemis at Ephesus one night in 356 BC. His only motive would have been to draw attention to himself, and in this he succeeded admirably, for although the Ephesians swore that his name was never to be uttered again, it continues to be mentioned to the present day, especially in the literature of psychiatry and psychoanalysis.[31]

Fire in Rome is a subject about which we have better documentation. This is not surprising, given the fact that the city grew to a size unequalled anywhere in Europe in its time. Because of its sheer size, as well as because of the poverty in which most of its people lived, it has often been compared to a Third World city in the twentieth century. As in Third World cities today, fire hazards were great, and blazes occurred frequently. Thus for the period between 31 BC and AD 410, contemporary authors recorded no fewer than forty large conflagrations in which numerous public buildings and large residential districts were destroyed – an average of one such disaster every eleven years. The number of smaller fires can only be guessed at.[32]

Eight of the forty conflagrations occurred during the Augustine period, between 31 BC and AD 14. By this time, Rome appears to have grown to a population of almost a million.[33] In spite of zoning regulations, tenement houses were built very close to each other. Their construction contained a great deal of timber and 'wattle and daub', which, according to Vitruvius, should never have been invented: 'For it is made to catch fire, like torches.'[34] As the satirical poet Juvenal was to remark, life in Rome's urban heart resembled an 'endless nightmare of fire and collapsing houses' (3: 6–8). Therefore, the poet continued,

I prefer to live where fires and midnight panics are not quite such common events. By the time the smoke is got up to your third-floor apartment (and you are still asleep), your heroic downstairs neighbour is roaring for water, and shifting his bits and pieces to safety. If the alarm goes at ground-level, the last to fry will be the attic tenant, way up among the nesting pigeons with nothing but tiles between himself and the weather. (3:198–203)[35]

As early as 450 BC there were city regulations stipulating, among other things, that houses should not be built too high or too close to each other, with more than $2\frac{1}{2}$ feet between them. It seems that from the start these rules were repeatedly ignored and therefore reissued. Augustus, reminding the Senate of ordinances of 105 BC, set the maximum height of houses at 70 feet, which again led to violations.[36] Later, in Imperial times, it was made compulsory to provide *insulae* with covered galleries at street level, which could serve as a means of escape. From early on, so it seems, tenants were required always to have a bucket of water in their flat. Rental contracts sometimes contained clauses forbidding the renter to make any open fire, on penalty of expulsion.[37]

In the days of the Republic, Rome had an official fire brigade (a *familia publica*) of slaves under the command of chosen members of the senatorial class, but apparently it was increasingly unable to cope with its task. This left the way open for enterprising politicians and businessmen to set up fire brigades of their own, organized as a *familia privata*. Thus we are told how Marcus Crassus (115–53 BC), a self-made man, started his career as the commander of a corps of 500 slaves trained as builders. In the event of a fire, he would appear on the scene with his men, and he would buy, at a trifling price, the burning houses and the houses adjoining them. He would then have his men put an end to the fire (probably by tearing down the adjoining houses), and immediately start rebuilding tenements to let or sell at a large profit. 'In this way,' the Greek historian Plutarch, in his *Lives*, wrote of Crassus, 'the largest part of Rome came into his possession.'[38]

A similar story was told about the politician Egnatius Rufus, who, during the latter days of the Republic, had raised a private fire brigade, and used it not to enrich himself financially but to gain political support. It was partly in order to forestall his rise to power that in AD 6, after another serious outbreak of fire, Augustus reorganized the existing public fire brigade into a new corps with a new name, the *Vigiles*.[39]

According to the ancient historian J. S. Rainbird, the *Vigiles* consisted of seven cohorts of nominally 560 men each; this number was doubled in AD 205. The reason for such a high concentration of manpower was, as Rainbird notes, the limited technology available. Intensive patrols at night were performed in order to discover fires while they were still small, and therefore easy to extinguish. As equipment the patrols carried buckets and axes; since they did not have hoses, their first action on discovering a fire would be to form a chain of men for passing buckets filled with water from the nearest reservoir. Clearly, this method would be effective only if the fire was discovered while it was still small.[40]

Although diminishing in frequency, big conflagrations continued

to occur. The largest, and best known, took place in AD 64, during
the reign of the emperor Nero. It raged for nine days, and in that
time almost a third of the city was destroyed by the fire or as a result
of attempts – mostly unsuccessful – to stop it. The event left such a
deep impression that several Roman historians, including Tacitus,
Suetonius and Diodorus Cassius, wrote about it. According to the
latter two, the fire had been lit by Nero in order to create space for
new palaces and parks. It is impossible to check the veracity of these
allegations, but Nero was not the only emperor in history to be
accused of having caused a conflagration in his capital; many Mus-
covites, for instance, held Ivan the Terrible responsible for the great
fire of 1550. I would not be surprised if there were more such stories,
testifying to a popular belief in the depravity of despots as well as
their omnipotence.

The ancient historian R. F. Newbold estimates that in the fire of
AD 64 at least 10,000 to 12,000 *insulae* were destroyed, plus several
hundred *domi*, leaving more than 200,000 people homeless.[41] Large
sections of the city had to be rebuilt, at a pace that could only be
detrimental to quality. Meanwhile, in the quarters that had been
spared, house and room rents went up steeply, forcing many tenants
and subtenants to leave and join the homeless. Although some owners
and landlords saw the value of their property and their income rise
fabulously, most people were badly duped. As a first measure, the
Imperial treasury was drawn upon to render assistance. This unleashed
a chain reaction, for, in accordance with the prevailing power
relations, the costs were subsequently passed on to the countryside.
Eventually it was the farmers in the provinces who had to bear the
brunt of the financial consequences of the fire of Rome – just as, not
infrequently, in times of famine the situation in rural areas was worse
than in the cities; especially in bad times, the tentacles of the cities
reached far and deep into the countryside.[42]

In the absence of any insurance schemes, people who lost property
in a conflagration could only hope for compensation by means of
gifts. Here again one's place in the social hierarchy was decisive. It
was said of slaves that they had nothing to lose but their lives, whereas
rich people could turn into paupers overnight if their property was

burned down or looted by criminals taking advantage of a fire.[43] With regard to the wealthy, however, Juvenal took a different view: 'If some millionaire's mansion is gutted ... contributions pour in while the shell is still ash-hot' (3: 216–18). It was certainly true that victims of a blaze were dependent on the benevolence of their family and friends; in the event of a huge catastrophe, however, help could come only from the highest chief of the *familia publica*, the emperor. Thus, because it gave him an opportunity for displaying his generosity, a big fire sometimes helped to enhance an emperor's popularity.[44]

Compared with the information on the city of Rome, we know very little about fires and fire-prevention in all the provinces of the Roman Empire. We do, however, have two interesting letters written in AD 112 by the Younger Pliny and the Emperor Trajan, both recorded in *The Letters of the Younger Pliny*. The Younger Pliny was at the time a provincial governor in Asia Minor. At issue was a matter raised by Pliny.

While I was visiting another part of the province, a widespread fire broke out in Nicomedia which destroyed many private houses and also two public buildings (the Elder Citizens' Club and the Temple of Isis) although a road runs between them. It was fanned by the strong breeze in the early stages, but it would not have spread so far but for the apathy of the populace; for it is generally agreed that people stood watching the disaster without bestirring themselves to do anything to stop it. Apart from this, there is not a single fire engine anywhere in the town, not a bucket nor any apparatus for fighting a fire. These will now be provided on my instructions.

Will you, Sir, consider whether you think a company of firemen might be formed, limited to 150 members? I will see that no one shall be admitted who is not genuinely a fireman, and that the privileges granted shall not be abused: it will not be difficult to keep such numbers under observation. (10: 33)[45]

To a modern reader, the emperor's reply may come as a surprise.

You may well have had the idea that it should be possible to form a company of firemen at Nicomedia on the model of those existing elsewhere, but we must remember that it is societies like these which have been responsible for the political disturbances in your province, particularly in its towns. If people assemble for a common purpose, whatever name we give them and for whatever reason, they soon turn into a political club. It is a better policy then to provide the equipment necessary for dealing with fires, and to instruct property owners to make use of it, calling on the help of the crowds which collect if they find it necessary. (10: 34)

A world of very unequal and, at the same time, precarious power relationships appears to open up in this brief correspondence. In our own time, some nightmarish dramas and novels have been written about firemen exploiting their special expertise not to extinguish but to light fires, and thus to terrorize people.[46] There are a few hints in classical historiography that in antiquity, too, firemen were sometimes seen as potential arsonists. Thus Tacitus reported the rumour that the *Vigiles* had been instrumental in spreading the great fire of AD 64. In later periods firemen were occasionally suspected of committing arson, for a variety of motives ranging from looting to self-aggrandise-ment and 'pyromania'. Yet, the perennial problem of who should guard the guardians seems seldom to have been raised with regard to fire brigades. And when Trajan refused his permission to set up a fire brigade in Nicomedia, he was not particularly concerned about arson. He feared only that people officially organized as a fire brigade would use their assemblies as a pretext for subversive political activities.

This is not to say that incendiarism did not worry people, including the authorities. Slaves could, and no doubt often did, cause fires through various grades of 'negligence', ranging from inattention to downright sabotage.[47] The ever-present risk of arson, in combination with the lack of any insurance schemes, may have been one of the conditions which discouraged Roman estate managers from making great capital investments. Setting fire to the property of landowners and their representatives was a regular feature during popular upris-ings.[48] The authorities, on their part, also used this weapon; thus tax collectors in the provinces would burn down the houses of peasants who refused to pay.[49]

FIRE IN RELIGION

In their religious practices the ancient Greeks and Romans differed greatly from the ancient Israelites. They had no sacred books; nor did they have a class of professional priests and prophets. They honoured many gods and they held numerous festivals throughout the year, with a great many local variations. Yet for them, as for the ancient Israelites – and for most other people in agrarian societies – the central act of worship was sacrifice, and most sacrifices were associated with meals and were made on a burning altar.

As noted by Fustel de Coulanges, the master of a Greek or Roman household would attach great importance to the domestic altar. A sense of that importance still resounds in the way historians write about it today; thus they will say, 'The Greek household had its shrine to Hestia or to Zeus Ktesion, either of whom could give special protection to hearth and home.'[50] I doubt whether any modern historian seriously believes that either Hestia or Zeus 'could give special protection', but this is the customary (and conveniently simple) way of writing about such matters. The crucial point is, it seems to me, that for a long time masters of a household used to perform a daily ritual in front of the hearth; and that this moment of piety might help to remind them of the preciousness of their social status and their possessions, of the precariousness of all this and of the obligations and responsibilities it implied. Viewed in this light, the sacrificial act appears as a constituent of the civilizing process, re-enacted every day.

According to Fustel de Coulanges, the altar fire was always to remain pure. This meant that 'no blameworthy deed ought to be committed in its presence'.[51] A linguistic purist might object that no blameworthy deed ought to be committed anywhere, but again, the tenor is clear: the fire was treated with veneration, and a reminder of its presence could cause people to refrain from doing something wrong. Fustel de Coulanges was probably right in suggesting that the custom of keeping the fire always burning was a relict of ancient times, when there were only a few fires available and great care had to be taken that they did not go out.

An intriguing aspect of the hearth cult is that, while it was performed by the master of the house, the worship was formally directed
at Hestia, a goddess. Hephaestus, the male fire god, was associated
with the world of smithies and battlefields. Hestia, for her part,
symbolized the unity and continuity of the household. Among her
attributes were chastity and peacefulness; she appeared to represent
the 'quiet' virtues of the ideal wife, faithful and obedient to her
husband.[52] Even in the war-torn world of the *Iliad* and the *Odyssey*,
she never took part in combat.

Compared with Hephaestus, as well as with the demigod Prometheus, Hestia played an unadventurous role in Greek mythology.
This seems to reflect the fact that the activities centred around the
domestic fire tended to be uneventful. Hestia's femininity also suggests
that the actual responsibility for keeping the home fire burning lay
with the mistress of the house. If there were no slaves in the household, she would be the one to attend the fire. As manual work,
carried out by women or slaves, this task would not have ranked
highly.

Plato, who regarded the strongly competitive spirit of Greek men
as a serious threat to the functioning of the *polis*, seems to have been
the first to express a more positive appreciation of Hestia. In the *Laws*
(745b, 848d) he put her on a par with Zeus and Athena; the main
temple in Plato's ideal city would be dedicated to this divine trio.
This, however, was never to be more than a wishful thought. While
Hestia was a generally respected goddess, her place on Olympus
remained modest, and there were no public festivities devoted to her.

As a public equivalent to the domestic hearth fire, Greek cities
used to have a *prytaneum*, a temple in which the 'sacred city fire'
would be kept burning day and night. It seems that Greeks would
always take fire from their *prytaneum* with them when they went out
to found a new colony. The Dutch classicist Lyda Simons suggests
that the *prytaneum* was the cultic centre of a local political community;
as such it differed from the few supra-local temples, like the temple
dedicated to Apollo at Delphi, in which a fire was kept burning
which symbolized a sense of national unity of a religious rather than
a political nature.[53]

Rome, by contrast, had in its city centre a temple fire dedicated to Vesta which signified both the religious and the political unity – at first of the Republic and later of the Empire. Like all temple fires, the Vestal fire served purely ceremonial functions. It was tended by a few (four at first, and later six) unmarried girls and women – the Vestal Virgins – who were appointed at an early age to be priestesses for thirty years. The Vestal Virgins used to belong to patrician families, and just as the female members of a household had to obey the *pater familias*, so they were subject to the authority of the supreme priest, the *pontifex maximus*. They had to live in accordance with strict rules. Violations were punished harshly; thus if a Virgin was found to have broken her vow of chastity, she was buried alive.

The official duty of the Vestal Virgins was to see to it that the fire in their temple was kept burning. The great importance attached to this task may have been a reflection, partly, of chauvinism and, partly, of the need to counteract the numerous centrifugal tendencies that were continuously at work in the realm. The cult of the Vestal Virgins, who, as unmarried women, were supposed not to be involved in party factions, was one of the centrally binding institutions.[54]

Of course, an altar fire was kept burning not just in the temple of Vesta. While the functions of such fires were mostly ritual, they could serve other purposes too. Thus for a long time, especially in Greece, temples along the coast would serve as landmarks for seamen which, at night, would be visible by their fires. From the first century AD, however, towers were built without any sacred functions just to serve as lighthouses. The famous Pharos at Alexandria, one of the seven wonders of the ancient world, was also converted for this purpose; at night, fires were burnt in its lanterns so that it could be seen from a distance of up to 30 miles. According to the Dutch historian of technology R. J. Forbes, 'By the end of the first century AD there must have been some thirty lighthouses, which number was slowly increased during the following century.'[55] The building of purely 'secular' towers to serve as lighthouses and thereby to take over one of the traditional functions of temples may be seen as indicative of a process of increasing specialization leading to secularization.

Meanwhile, many local cults persisted. The Greek author

Pausanias, in his *Guide to Greece*, written in the second half of the second century AD, found occasion to describe a great variety of religious practices and ceremonies. Thus he related how the people of Patrai celebrated a yearly festival dedicated to Artemis. By way of preparation, they would build a large pyre round the altar, using logs with a length of 30 feet. On the day of the festival

> they throw game-birds live on to the altar and all the other victims in the same way, even boars and deer and gazelles, and some of them throw on wolf-cubs or bear-cubs and others fully grown beasts, and they heap the altar with fruit from orchard trees. After this they set fire to the logs. At this point I saw a bear and other beasts forced out by the first leap of the flames or escaping at full strength; but those who threw them in bring them back again to the funeral fire. They have no record of anyone being injured by the animals. (7.18.7)[56]

The entire scene, in which human mastery over both animals and fire was made clearly manifest, was described by Pausanias without any hint of disapproval or abhorrence. He only noted that 'The whole city as a body prides itself over this festival, and so do the individuals just as well.' Indeed, the proceedings seem to have been very orderly, and the level of cruelty to the animals did not exceed that displayed almost routinely in the public games in the capital city of Rome.

Throughout the empire the Imperial cult had to compete not only with persisting local cults but also with some rising oriental religions; one was Mithraism, an esoteric cult centred on the sun, and another was Christianity. At the time of the great fire of AD 64, Christians were still regarded as a pernicious sect, whose members could be used as scapegoats when rumours started to spread that the emperor himself had started the conflagration. In his *Annals* on the history of Rome, Tacitus related how they were arrested and convicted,

> not so much on the count of the arson as for hatred of the human race. And derision accompanied their end: they were covered with wild beasts' skin and torn to death by dogs; or they were fastened on crosses, and, when daylight failed, were burned to serve as lamps by night. (15:14)[57]

Christianity had begun, as Tacitus knew, in Palestine, with the teachings of an Israelite known as Christ. Acting in the time-honoured tradition of the Jewish prophets, Christ moved, further than any of his predecessors, away from the ritualistic adherence to an agrarian regime as represented in Leviticus and Deuteronomy. Instead, he addressed the moral issues with which people living in a complex urban-agrarian world saw themselves confronted. When he made reference, as he often did, to agrarian practices, it was by way of parable, not with a view to instructing farmers about their work. Thus, according to the gospel of Matthew, in order to explain the idea of the kingdom of heaven to his followers, he told them about a man who sowed good seed in his field and how, one night, his enemy came and sowed weeds among the wheat. When the wheat came up, so did the weeds. The man's servants then suggested that they should go and gather the weeds, but the man wisely answered that this would be impractical, for they would root up the wheat along with the weeds. He ordered them to wait until harvest time; then they would be able to gather the weeds first and bind them in bundles to be burned, so that the wheat could be reaped and put safely into the barn (Matthew 13: 24–30). When, somewhat later, his disciples asked Christ to explain this parable, he answered:

He that soweth the good seed is the Son of man;
The field is the world; the good seed are the children of the kingdom; but the tares are the children of the wicked one;
The enemy that sowed them is the devil; the harvest is the end of the world; and the reapers are the angels.
As therefore the tares are gathered and burned in the fire; so shall it be in the end of this world.
The Son of man shall send forth his angels, and they shall gather out of his kingdom all things that offend, and them which do iniquity;
And shall cast them into a furnace of fire: there shall be wailing and gnashing of teeth.
Then shall the righteous shine forth as the sun in the kingdom of their Father. Who hath ears to hear, let him hear. (13: 37–43)

This highly suggestive parable echoes in many other passages in the New Testament, the collection of documents dating back to the first century AD which present an account of the life and teachings of Jesus Christ and his immediate followers. A recurrent theme in these Scriptures, continuing the tradition of moral prophets such as Jeremiah, is concern for personal salvation, for the well-being of one's soul. People are warned never to indulge in anything sinful; even if in this life they are able to get away with a crime, there will be no escape from punishment in the life hereafter. Persons who commit adultery or theft will bitterly regret any enjoyment it may have given them, for they will pay for it with eternal damnation. On the day of judgement, they will be thrown out with 'the chaff that will burn with unquenchable fire' (Matthew 3: 12, Luke 3: 17). Therefore, Christ is reported to have said:

Woe unto the world because of offences! for it must needs be that offences come; but woe to that man by whom the offence cometh!

Wherefore if thy hand or thy foot offend thee, cut them off, and cast them from thee: it is better for thee to enter into life halt or maimed, rather than having two hands or two feet to be cast into everlasting fire.

And if thine eye offend thee, pluck it out, and cast it from thee: it is better for thee to enter into life with one eye, rather than having two eyes to be cast into hell fire. (Matthew 18: 7–9; cf. Mark 9: 43–8)

These gruesome warnings find a climax in the final book of the New Testament, the Revelation of Saint John. Among other apocalyptic prophecies, it contains some obscure but terrifying visions of fire coming down from heaven and consuming people, and of an eternally burning lake of fire and sulphur (8: 5–8; 16: 8–9; 21: 8). Dreadful though these visions are, however, we should not forget that they are only parts of a more general picture of torment and chaos, drawn up as the opposite of the kingdom of heaven. In the apocalypse, fire was just one of many ordeals.

The teachings in the New Testament may be considered as a very eloquent 'civilizing campaign'. In these teachings an appeal was made to people to exert greater self-restraint in their dealings with each other – an attempt to strengthen the voice of conscience within them. The texts were not written with great confidence in people's capacity to act decently of their own accord. The appeal to inner restraint was continually reinforced by reference to the external authority of a god who, at the day of judgement, would sift the wheat from the chaff, and consign the chaff to the flames.

Perhaps the vision of a perpetual fire in which the souls of the damned were to burn for ever was inspired, to some extent, by the experience of city fires, especially in Rome. The heat and the stench of a multitude of industrial furnaces and of long-smouldering rubbish heaps may also have contributed to the idea of infernal fire. Be that as it may, the theme of the horrors of fire was available to the moralizing imagination.

Christianity, which during the reign of Nero was still the religion of a small minority, grew rapidly in numbers and in influence. At the end of the third century AD the Emperor Diocletian made a last, desperate effort to put an end to it, but he failed, and in 312 his successor, Constantine, officially espoused Christianity. The Visigoths, who in 410 sacked the city of Rome, were Christians; Saint Augustine thought it greatly to their credit that they spared the lives of all those who had sought refuge in a church.

FUEL AND DEFORESTATION

The demise of the Roman Empire, certainly the Western part, was marked by a reversal of the trends that had accompanied its growth. Populations shrank, cities declined or even disappeared altogether (Rome itself diminished to the size of a provincial town), leaving less anchorage for specialized crafts and trades and for large-scale governmental and military organizations. Demographic decline, in short, went together with economic and political disintegration.

It has been suggested more than once that a major cause of this collapse was the depletion of forests. The ancient historians J. Donald Hughes and J. V. Thirgood have stated the case strongly.

Forests provided the major material for construction and almost the only fuel source of the classical world, and depletion of this source precipitated a number of crises. As forests retreated with land clearance, wood decreased in availability and increased in price, contributing to the ruinous inflation that plagued antiquity. Competition for forest resources ignited military conflicts, which themselves created demands for timber. Erosion weakened the economic base of the predominantly agrarian societies, contributing to a population decline that made it ever more difficult for Graeco-Roman civilization to resist the incursions of barbarians from beyond the frontiers. In the more arid regions, forests that formerly moderated the climate and equalized the water supply were stripped away, permitting the desert to advance. The image of the ruined cities of North Africa, from which olive oil and timber were exported in ancient times but which later were buried beneath the desert sand, epitomizes the environmental factor in the decline of civilization, as do the swamps along the northern Mediterranean margin from which malaria spread to debilitate the population.[58]

The argument is far-reaching and stated in persuasive terms, and may find support in some ancient literary sources. Thus as early as the fourth century BC Plato indicated the barrenness of the region surrounding Athens.[59] There are a few more testimonies to similar concerns, but on the whole the writers of Greek and Roman antiquity evinced complacency rather than disquiet about human mastery over nature. If, indeed, there was an increasing shortage of wood, this does not seem to have seriously affected the privileged circles to which they belonged.

For many centuries, as the ancient historian Russell Meiggs observes, the history of the Roman world was marked by a growing population and a rising standard of living. Increasingly more timber was used for building, increasingly more fuel for cooking, for industrial purposes, and for heating houses and public buildings, including the numerous vast public baths. Wood could not fail to become scarce, and yet, Meiggs notes, 'there is no evidence of any general alarm at the depletion of the forests and there is no evidence of any attempts to redress the balance'.[60]

Hughes and Thirgood do mention some attempts at conservation, such as protection by local magistrates of certain woods designated as sacred groves. These measures were far outweighed, however, by the continuous pressure from people eager to convert forest into arable land and to sell wood on the market for fuel and timber. Moreover, 'the constant warfare that, except for a few happy decades, afflicted the Mediterranean world' also took its toll.[61] Even in peacetime the army was a prodigious consumer of fuel; in times of war, consumption rose even higher, and to it was added deliberate destruction.

All in all, deforestation was but one in a pattern of interconnected processes. It was the consequence of many forces, and at the same time a cause of many others. Subsequently, in those parts of Western Europe where erosion had not yet made the depletion of forests irreversible, the long period of disintegration of the Roman Empire allowed for a spontaneous process of reforestation.

7. FIRE IN PRE-INDUSTRIAL EUROPE

THE FOUR ESTATES

The development of the control of fire may serve to remind us of the continuity and the unity in human history – two aspects that tend to be overshadowed by the more spectacular turning points and unique features that form the hallmarks of distinct eras and cultures.

The emphasis on continuity in the control of fire is by no means intended to suggest that its development was either unilinear or without shocks. Each conflagration formed an interruption, each technical innovation, no matter how minor, gave new impulses. In retrospect, however, we can see how, over the centuries, all these events were interlocking and how they constituted one single process.

This process encompasses different historical periods, as well as seemingly unrelated cultures. In this chapter, I will focus on Western Europe between 850 and 1850 – roughly the same area and era that Norbert Elias covered in *The Civilizing Process*. During this age, society in Western Europe slowly began to deviate from the military-agrarian pattern prevalent throughout most of Eurasia. From the very start, however, this specific development occurred in contact with other regions, as part of a wider configuration. During the first five centuries, the Western Europeans were mostly at the receiving end of a chain of cultural diffusion, learning a great deal from Byzantines, Arabs, Indians and Chinese. But then the tables were gradually turned, with Europe exerting increasing influence on the rest of the world.

In these world-historical shifts of cultural influence and social dominance, the control of fire did not play an independent part. Yet the ways in which people managed to increase their means of exploiting fire, and to develop more efficient strategies to protect their property against it, had far-reaching ramifications that are well worth exploring.

In *The Civilizing Process* Elias highlighted the connection between the civilizing process and state formation in Europe, with the social pressures arising from the interdependence of progressively larger groups of people as the connecting link. Increasing interdependence also played a central part in the development of the control of fire. Unlike the refinements of etiquette, however, which formed the point of departure for Elias's study, changes in the control of fire did not originate at the princely courts but elsewhere in society. Accordingly, in this chapter I will follow a different route from that which Elias pursued in his book. I will not focus on the aristocracy, but instead will look, in consecutive sections, at each of the four 'estates' which formed the major sections of society in pre-industrial Europe: the clergy, the nobility, the bourgeoisie and the peasantry.

Needless to say, the four estates were interdependent. During the long agrarian phase of European history, the differences in conduct and power between them appeared to become more pronounced. At the same time, however – as the great French sociologist Alexis de Tocqueville was perhaps the first to perceive – there was from early on an undercurrent of convergence.[1] Some of the great turning points in this era, such as the Reformation and the French Revolution, can be seen as manifestations of this counter-trend towards diminishing differences in conduct and power between the four estates.

Each of the estates lived under particular social constraints, and developed its own lifestyle accordingly, partly in response to its position *vis-à-vis* the other estates. Not surprisingly in this situation, different views came to be articulated as to where the major agencies of the European civilizing process were to be found. Saint Augustine, and other clerical writers following in his footsteps, were the first to address this issue, attributing the most important civilizing impetus to the institutions of the 'first estate': the church, the monasteries, religion. A later intellectual tradition, of which Thomas Hobbes was one of the most prominent representatives, gave priority to politics and the state – institutions originally dominated by the 'second estate': the nobility. Later again, ideologists of the bourgeoisie such as Adam Smith pointed to the civilizing functions of the institutions of the 'third estate', notably the market and the prosperity and peace it generated. Only the agrarian regime of peasants and farmers was seldom singled out as a major force in the European civilizing process.

FIRE AND RELIGION

Religion, Fire and the Narcissism of Small Differences

One of the seemingly self-evident peculiarities of social and cultural development over the past 2,000 years is the predominant role of large-scale organized religions. It seems not to call for any further comment to say that Western Europe, ever since the early Middle Ages, belonged to the Christian world. Religion apparently has become a decisive element in people's identity.

This is very different from the situation in the ancient world. The religion of the Egyptians was named after Egypt; one would not name the Egyptians after their religion as 'Raists', nor the Romans as 'Jupiterians'. The change that has occurred points to a considerable extension of the range of influence of religious regimes.

The rise to predominance of organized religion is closely related to the collapse of the Roman Empire. As noted by Judith Herrin, the British historian of Byzantium, the break-up of the Empire eventually led to the formation of three socio-cultural areas within its former territory: the world of the Greek Orthodox Church, dominated by Byzantium; the world of Western Christianity, committed to the Roman Catholic Church; and the Muslim world. It was a tripartition in which religion appeared to be of paramount importance.[2]

The preoccupation with religious purity which had been typical of ancient Judaism was taken over by the new religious regimes. Like the priests and prophets of Judaism, the Christian and Muslim leaders laid great stress on monotheism, and condemned the worship of idols. Among their official orthodoxies was also a rejection of fire cults, such as the worship of the fire god Atar in the Parsee or Zoroastrian religion, and the worship of Agni in Hinduism.[3]

Perhaps the negative attitude towards fire cults had something to do with the fact that the priestly élites who were highly influential in the formation of Judaic, Christian and Muslim dogma lived in the urban world of such cities as Jerusalem, Rome and Baghdad. None of these cities was very well suited to the staging of fire festivals; moreover, in view of the advanced division of labour, the priests had no chance of competing with such specialists as smiths in mastery over fire.

While there is some plausibility in each of these reasons for the rejection of fire cults, I think that they should be seen in combination with another principle which always seems to play a part in the dynamics of religious organizations – the principle known in the literature of psychoanalysis as the 'narcissism of small differences'.[4]

As the adherents of a particular creed, theologians usually tend to emphasize those elements which make their own religion special and exclusive. By doing so they actually contribute to making their religion special. It is a good example of how a successful 'definition of the situation' may indeed affect the situation so defined. Religions differ from each other, partly because those who are 'in charge' set great store by those very differences. To a considerable extent religious specialists owe their position to the distinctions between their own religion and any possible rival cults and creeds. They are therefore likely to cultivate the features which they consider to be characteristic of their own religion and strictly to reject every deviation. To do otherwise would run counter to their interest.

This principle may help to explain why the Fathers of the Church in their official doctrine discouraged fire worship, to make a clear distinction between Christianity and the initially competing cults of Mithras and Zoroaster. Yet, from the start, Christian worship was full of the symbolism of fire and light.[5] The Church in Western Europe adopted several 'heathen' customs with fire, both from the Romans and, as they were converted, from the Celtic and Germanic peoples.

Moreover, it developed some spectacular practices and beliefs of its own in which fire played a crucial part.

Fire Festivals

During the first centuries following the collapse of the western Roman Empire, the dominant trend in the political and military structure of Western Europe was towards decentralization. At the same time, however, centralization occurred in the religious sphere. Building upon the remnants of the cultural and material infrastructure of the Empire (vestiges of 'intensive growth'), the Church of Rome gradually set up a tight organization by means of which it brought some uniformity in religious life over a wide area. In all temples (or churches as they were now commonly called) priests observed basically the same liturgy. In this liturgy fire played a modest but essential part. Like the ancient Roman temples, all Christian houses of worship contained a perpetually burning altar fire – as a tacit reminder of the times when fire was a precious group possession. The fire in the church did not need to be more than a simple small lamp that would function during the services as a symbol of light, not of warmth or scourging heat.[5] Sometimes, on feast days such as Easter or Christmas, the churches were lit up brilliantly. On ordinary days, the scent of burning incense contributed to the sacred atmosphere.

On special occasions there were also open-air fire festivals in which people would kindle large bonfires or hold processions or races with blazing torches round fields or through village streets.[6] Although these festivals had no direct connection with Christian beliefs, they became, in the course of time, incorporated into the Christian calendar, and associated with Christian saints. According to the German folklorist Herbert Freudenthal, these festivals expressed the ambivalent feelings of the rural population towards fire. People feared its destructive powers, but they also cherished it for creating warmth and light, and regarded it as a faithful ally against numerous dangers which threatened their existence. They used it to destroy refuse and, with some imagination, other less tangible evil as well: 'to burn or repel', as Sir James Frazer put it, 'the noxious things, whether conceived as material or spiritual, which threaten the life of man, of animals, and of plants'.[7] Folklorists tend to emphasize the magico-religious aspects of these purification rites. It is worth noting, however, that for a rural community with a relatively low level of fuel consumption, a fire festival could also have the more practical purpose, at the end of a season, of getting rid of waste matter that, if left lying about, might present a serious fire hazard.

One of the largest fire festivals was the midsummer feast, tied, since Christianization, to Saint John's Day. For a long time, it was celebrated both in the country and in cities. Great piles of combustible material would be erected on village greens or town squares, and lit on the proper evening at sunset. Rubbish that had been collected over many months now went up in flames. To heighten the suspense people might throw puppets ('straw men') on the fire, or, as was customary in many places, living animals. Thus in early modern times Saint John's Day was celebrated in Paris with a bonfire in which cats were burned alive. Frazer summarized the proceedings as follows.

In the midsummer fires formerly kindled on the Place de Grève at Paris it was the custom to burn a basket, barrel or sack full of live cats, which was hung from a tall mast in the midst of the bonfire; sometimes a fox was burned. The people collected the embers and ashes of the fire and took them home, believing that they brought good luck. The French kings often witnessed these spectacles and even lit the bonfire with their own hands. In 1648 Louis XIV, crowned with a wreath of roses and carrying a bunch of roses in his hand, kindled the fire, danced at it and partook of the banquet afterwards in the town hall.[8]

In *The Civilizing Process*, Norbert Elias also discussed the Saint John's festival in Paris. The cat burnings, which arouse revulsion in us, must for many generations have caused an enjoyable sensation — like a football game or a boxing match today. This shows, according to Elias, the degree to which feelings of pleasure and displeasure are subject to changing social standards. The 'great cat massacre' continued to be performed in Paris each year until far into the eighteenth century, although the king and the nobility no longer took part in it.[9] The festival had socially degraded into a 'popular event' in a more modern and restricted sense of the word — an item of culture that had 'sunk down' and from which the élite stayed aloof.

In Germany similar developments took place, having started perhaps slightly earlier. Thus Freudenthal (who made no mention of any burning of cats) noted that already in the early fifteenth century some town governments were trying to put an end to the Saint John's bonfire, probably because of the fire hazard and the general disorder accompanying it. In spite of these attempts, at the end of the century the festival was still widely celebrated and people of the highest ranks, including kings, took part in it. In the sixteenth century, however, an avalanche of prohibitions broke out, by means of which worldly and clerical authorities jointly tried to abolish the Saint John's bonfires. In the long run this 'civilizing campaign' was effective, and by the beginning of the nineteenth century it was only in the country that this fire festival was still celebrated; and even there it was increasingly confined to the lower social classes and, eventually, to youth.[10]

The twentieth century saw a renewed interest in Saint John's festival, occasioned by the same mixture of nationalistic and commercial motives that has caused the revival (or, equally often, the 'invention') of many forms of folklore.[11] There is thus a clear pattern in the overall development: first, from purification rites celebrated, with the approval and active participation of Church and state authorities, by the entire community, to forms of popular amusement, mainly for the young; and then, in our own time, the trend towards a 'revival' in the guise of more respectable events sponsored by businesses and politicians.

Stakes and Visions of Hell

Some centuries after the Christian Church had definitely established its authority in Western Europe, fire was once again given a more important role in religion. Its terrifying effect was particularly stressed, in two ways: in the form of stakes at which people who had been accused of heresy or witchcraft were brought to their death, and in visions of hell and purgatory.

This development poses at least two questions. First, how is it to be explained that the religious regime was tightened, leading to a new fanaticism which manifested itself in the persecution of 'heretics', 'witches' and other groups who were excommunicated from the community of believers? Second, why did the tightening religious regime rely so strongly on fire? Neither of these questions is easily answered, but something can be said about both of them.

With regard to the first, the British historian R. I. Moore makes some interesting observations in his book on the emergence of persecutions in Western Europe in the period between 950 and 1250. He sketches the common background of various persecuting movements: against heretics, Jews, lepers, sodomites and witches. Moore concludes that all these movements were part of a wider political struggle, related to the incipient consolidation of power in states. He notes something resembling a 'trickle effect'. The large-scale persecutions of witches in the sixteenth and seventeenth centuries, the victims of which were mainly women from the lower social orders, were the continuation of a process of demonization which had started with deadly intrigues within the upper strata.[12]

Clearly this is an ingenious interpretation, not the complete explanation. In answer to the second question, too, conjecture must suffice. We can note only that the horrors of fire were used to add force to the clerical campaigns – horrors in the form of physical torture at the stake and of mental torture through images of hell and purgatory. Probably the two kinds of agony were related, and the fear of hell and purgatory was reinforced by the public executions with fire.

In 1022 the king of France ordered fourteen members of the higher clergy and the leading citizenry in the city of Orléans to be burned as heretics. This brought to an end a period of more than six centuries during which, in Western Europe, there had been no charges of heresy followed by the death penalty. The victims in Orléans were the first in a long series. There were times, such as the period of the 'crusades' against the Albigensians in southern France in the early thirteenth century, when scores of people were burned at a time.[13]

It seems only too likely that the burning of heretics made people more prone to the fear of hell and purgatory. They knew that fire could be inflicted as a punishment; if they had not themselves witnessed an execution at the stake, they would have heard others talk about it who had. They were also familiar with the theological justification: burning served as a means of purification. How, then, could they put aside the idea that all sinful souls would be punished with fire, and that perhaps even all souls might be put to the test in an ordeal by fire?

The idea in itself had now been brought into circulation. The Bible contained only a few references, all in the New Testament, to a fiery hell awaiting every earthly sinner after death. These few passages, however, lent themselves to the most gruesome elaborations. The notion of purgatory was completely absent from the Bible; but when, around the middle of the twelfth century, it was first propagated, it proved to have wide appeal.[14]

According to Herbert Freudenthal, the idea of a hell filled with fire was at first wholly alien to the peoples of northern Europe; for them, the worst agony would have been an afterlife of eternal frost. The notion of a fiery hell could have originated only in a warm environment. There may be something in this climatological argument, but it does not help us solve the problem of exactly why, in the later Middle Ages, so many people in the north and west of Europe became obsessed by fear of a fiery hell.

Might contact with Islam have played a part? Whereas in the Bible references to hell-fire are scarce, they occur frequently in the Koran. Nevertheless, in spite of many allusions to 'the fire that has been prepared for the unbelievers' (Sura 2, 24), the Koran contains nothing to compare with the detailed accounts of torments by fire such as emerged in late medieval and early modern Christian Europe.

Whether in words or pictures, fire in hell always appeared in a domesticated guise. Devils were depicted as inflicting torture in ways not unlike those employed by executioners on earth. The infernal torments in the paintings of Hieronymus Bosch could have served as illustrations of the line in Jean-Paul Sartre's play *Closed Doors*: 'Hell is other people.' Even the fantasies of the seventeenth-century Italian Jesuit Father Pietro Pinamonti, evoking the very worst pains imaginable, were inspired by domesticated fire:

Every damned person will be like a heated oven, blazing hot on the outside and inside his chest; the filthy blood will boil in his veins, as will the brain in his skull, the heart in his chest, and the guts in his wretched body.[15]

The images of hell-fire as they were developed by successive generations of theologians, poets and painters reflected – as is true for all elements of culture – a measure of 'relative autonomy' that could not possibly be reduced to other factors.[16] Having recognized the validity of this principle, we may still be puzzled by the question of what constituted the social and psychological soil on which these images could thrive. In order to understand the civilizing campaigns of which the *autos de fe* formed a part, they need also to be seen in the wider context of a military-agrarian society with a steadily growing urban population. The people who feared fire in hell also had some knowledge of fire in war and fire in cities.

FIRE IN WAR

Greek Fire

The division of the Roman Empire into three distinct religious domains was not established by missionary persuasion alone; military force played a decisive part. A fearsome weapon in these military struggles, especially in the eastern Mediterranean, was, for a number of centuries, the so-called 'Greek fire'. The vague name (coined, so it seems, by the Christian crusaders from Western Europe), the secrecy by which it was surrounded and the many legends told about this weapon make it sometimes difficult to separate truth from fiction. Yet its history is worth examining.

According to legend, the Byzantine navy had, from the second half of the seventh century, a terrifying weapon with which to defeat any enemy at sea. Their ships were fitted with fire-throwing muzzles from which burning liquid could be shot at the enemy. The fire was not extinguished by the water, but continued burning. Any ship that was engulfed by it went up in flames.

The first to be confronted with this weapon were the Arabs. Their triumphal expansion along the borders of the Mediterranean was halted by it twice (in 676 and 717) in the waters near Constantinople. The rescue of the city, and therefore of the Empire (and, according to some, of Christianity), was generally attributed to the use of 'liquid fire'. In later centuries, several other naval powers – Franks, Russians, Normans, Pisans – were similarly deterred.

The Byzantines treated their weapon as a sacred treasure. In the tenth century, Constantine Porphyrogenitus, the ruling emperor, solemnly urged his son and heir never to reveal its secrets to any other nation:

We are assured by the faithful witnesses of our father and grandfather that it should be manufactured among the Christians only, and in the city ruled by them, and nowhere else at all, nor should it be sent and taught to any other nation whatsoever.[17]

The horrendous effects of Greek fire could not fail to elicit a great deal of speculation as to its nature, among contemporaries as well as among subsequent historians. The burning liquid was usually described as a mixture of sulphur, saltpetre, naphtha and other substances with sinister-sounding names, most of which suggested more precision than they actually conveyed. The traditional views have recently been questioned by historians of science and technology, who consider it more likely that the liquid consisted of a highly flammable variety of crude oil or a distillation of it. The 'secret' of its use rested, first, on knowing the location of the sources where the oil could be obtained, and on being able to gather it at a sufficiently low temperature so that it would not immediately evaporate. Then, during combat, devices and skills were needed to heat the oil in a sealed container and, with the aid of a pump, to project it on to the enemy as soon as it reached ignition. The whole operation was very hazardous and required highly skilled men; if they died in battle, they could be replaced only by equally well-trained specialists.[18]

For many centuries, the Byzantines claimed that they were the only power to possess this weapon. Strictly speaking, this was probably true. Even when, in 812, a Bulgarian force took by surprise an arsenal containing bronze tubes for the discharge of Greek fire, and a great supply of the liquid itself as well, they apparently lacked the skill to make any use of it. If indeed Greek fire continued to be a peculiarly Byzantine weapon, this would corroborate Martin van Creveld's view that warfare in pre-industrial society, before 1500, was marked, not by a single dominant military technology, but by 'a tendency towards regional and national specialization'.[19]

Thus at the same time the Arabs, using asbestos and other fire-resistant materials, developed increasingly effective armour and strategies to defend themselves against assaults by Greek fire. In addition, they invented fire grenades and other incendiary weapons of their own. It was this entire range of Arab and Byzantine weapons that the crusaders from Western Europe indiscriminately labelled 'Greek fire'.

For the crusaders, coming from regions without easily ignitable mineral liquids, the first encounter with the incendiary weapons developed in the Mediterranean area came as a horrible shock. In 1139 the second Lateran Council condemned the use of such weapons as a deadly sin. Yet, not long afterwards, armies of Christian soldiers from the West also tried to employ them; some of these attempts ended in dismal failures because the fire operators had misjudged the direction of the wind and set their own equipment alight.

After the twelfth century, the use of Greek fire and its equivalents seems to have diminished – not abruptly but gradually. Various reasons have been given for this. It may have been the result of the general weakening of the Byzantine state, causing a lack of trained operators and a loss of control over the supply lines for the necessary kind of oil.[20] Another reason why it fell into disuse may have been the limited possibilities of Greek fire. It could be fired only at close range, and was therefore effective only at sea where one could come sufficiently near to the enemy to make the burning liquid hit its target; even then the sea had to be calm and the wind reliable. Targets on land were neither as approachable nor as combustible as the wooden ships.

An alternative explanation for the gradual obsolescence of Greek fire may be that its deterrent qualities diminished as the Arabs and other military rivals came to possess equally effective incendiary weapons. Perhaps its destructive power in sea battles was so great that

fleet commanders were willing to take risks with it only against
enemies who were unable to retaliate in kind.

The eventual disappearance of Greek fire was undoubtedly hastened by the advent of a new sort of fire weapon: the cannon loaded
with gunpowder. Although initially also far from reliable, and far
from safe for those who used it, this type of weapon opened a
whole new range of opportunities for artillery. In 1453 the walls
of Constantinople crumbled under the projectiles fired by Turkish
cannon.

Even in their final defence, the Byzantines availed themselves of
incendiary weapons. They had to yield, however, to the force of
cannon, combined with harsh military discipline.

Fire was poured down on the Turkish troops who stormed the walls of the
city, and we are given a nightmare picture of the soldiers falling into the
moat screaming with pain, while more were beaten forward by guards with
maces and whips, and the Janitsars in the background cut down those who
fled. But in 1453 gunpowder was the decisive weapon, and Greek fire-
ships were sunk by Turkish cannon-balls before they could damage the
invading fleet.[21]

Gunpowder

In their time, the cannon which destroyed the walls and the ships of
Byzantium were highly advanced. They combined a heavy calibre
with a long range. Yet they could not stand comparison with later
artillery. Loading took a long time, aim was not very accurate and
every shot involved a risk that the entire hulk would explode. The
contraption was so bulky that it made transport practically impossible;
the weapons were therefore cast on the spot.[22]

All this may help to remind us once more of the extent to which
the value of technical attainments is always dependent upon the
'historical context'. Just as earlier the Byzantines had an edge on their
rivals thanks to Greek fire, now the Turks were superior by virtue of
their mammoth cannon. But this advantage, too, was only temporary.

Cannon were known, for good reasons, as fire-arms. They were
constructed on the principle that if a highly volatile compound could

be made to explode in a chamber, the force of the explosion would be capable of sending a projectile towards an enemy target. To make this principle work required both an appropriate chemical compound and a mechanical device in which the combustion was to take place.

The crux lay in the combination of the explosive substance (the 'powder') and a sturdy metal device (the 'gun') in which the explosion could take place. As early as the middle of the ninth century AD, as the British sinologist Joseph Needham points out, Chinese alchemists, while looking for such elusive substances as elixirs of life and material immortality, were learning to make highly explosive mixtures of saltpetre, carbon and sulphur. It was soon discovered that these mixtures could be brought to explode in bamboo cylinders, which were used as weapons, first as flame-throwers and later also to launch projectiles.[23]

There was something radically novel in utilizing explosions. Of old, the control of fire was directed primarily at producing a continuous and even process of combustion which could be used as a steadily available source of heat and light. Gunpowder, by contrast, caused brief and violent eruptions of energy. Once it became known in Western Europe, inventors soon began to look for means of regulating these eruptions in a manner which allowed the energy that they released to be used for propelling engines. For a long time, however, even the most brilliant minds, such as the Dutch scientist Christiaan Huygens, failed to solve this problem. It was not until the nineteenth century that 'automatic weapons', such as the machine-gun, in which a rapid and even sequence of explosions could be effected, were invented. Likewise in the nineteenth century, the same principle was applied for peaceful purposes with the internal combustion engine; here the fuel that was brought to rapid and regular explosions was, of course, not gunpowder but oil or petrol.

Apart from its application in fireworks, the uses of gunpowder remained almost exclusively military. The oldest references to such use date from the late thirteenth century in China, and the early fourteenth century in Europe.[24] It was to take at least another 100 years, however, before the destructive power of the cannon could equal that of cross-bows and catapults. The crucial advances occurred in metallurgy, as developments in casting and forging enabled armourers to produce guns with both increasingly greater calibre and precision. In the long run, no medieval fortress was capable of withstanding the new artillery. During an expedition in Normandy in 1449–50, the army of the French king, equipped with cannon, needed no more than one year and four days to force sixty castles to surrender.[25]

In Asia and eastern Europe military élites succeeded in establishing large empires with the aid of fire-arms. The ruling dynasties in these 'gunpowder empires', as William McNeill calls them, had little interest in further developing the arms industry.[26] Rather, they consolidated their position at the already existing level of military technology. This policy was carried to its furthest extremes in Japan. Not only did the government in that country establish, in 1588, an official monopoly on the possession of arms for the warrior class, the samurai (this measure had an equivalent in the gradual formation of state monopolies of organized violence in Western Europe), but it also managed virtually to abolish the use of fire-arms by the samurai themselves. On its islands, the dominant class was able to maintain itself by the sword without having to take up fire-arms.[27]

In Western Europe history took a very different course. Here an arms race started in which a flourishing war industry provided the artillery of competing armies with ever more effective cannon and the infantry with increasingly efficient hand guns and pistols. Advances in the technology of attack led to innovations in defence, such as earthen walls capable of intercepting cannon balls. One effect of the arms race was further to increase the superiority of the military over the rural civilian population. Peasants and villagers became more vulnerable than ever before to passing armies with their practices of trampling the fields, pillaging and fire-raising. It was partly in reaction to the suffering brought to the German countryside by the Thirty Years War that, in 1676–8, an international conference was held in order to check the terror imposed by armies over rural populations; according to the American historian Myron Gutmann, this conference had a 'civilizing influence'.[28]

The Western European arms race also led to the technique of

fixing cannon on ships, thus turning them into 'floating bastions'.[29] Just as the Byzantines had thought that they owed their power to the possession of Greek fire, so now Europeans tended to believe that the world hegemony they were beginning to establish was based upon fire-arms. As Robert Boyle, the famous British scientist, wrote in 1664: 'the poor Indians lookt upon the Spaniards as more than Men, because the knowledge they had of the Properties of Nitre, Sulphur and Charcoale duly mixt, enabled them to Thunder and Lighten so fatally, when they pleas'd.'[30]

Such contemporary comments probably overrated the part played by fire-arms in the first European conquest of America, and other parts of the world.[31] Hegemony was not attained by gunpowder alone. While surely in a great many situations the intimidating possession of fire-arms could well be decisive, their possession as such was based on a much larger configuration of political, economic and cultural conditions all of which contributed to the growth of military strength. Far from being an independent factor, the control of fire was, once again, embedded in the entire social structure.[32]

FIRE IN CITIES

Fire-prevention

Historians generally agree that the specific course of socio-cultural development in Western Europe since the early Middle Ages was manifested, more than anywhere else, in the cities. Opinions seem to differ, however, about the exact nature of the relationship between the cities and the country. Thus the Italian historian Carlo M. Cipolla writes in his introduction to the *Fontana Economic History of Europe*: 'In medieval Europe, the town began to represent an abnormal growth, a peculiar body totally foreign to the surrounding environment.' His French colleague Jacques Le Goff, on the other hand, in his contribution to the same volume, stresses how deeply 'the medieval city was impregnated with the country'.[33]

These apparently contradictory statements may perhaps be reconciled by pointing to the fact that medieval Western Europe was characterized by a proliferation of relatively small towns, and the continued absence of any large metropolises. The cities indeed stood out as distinct entities in the physical and social landscape; but they did not develop into colossal urban conglomerations on a par with ancient Rome or Constantinople.

As in the ancient world, the concentration of people, property and fires made cities highly vulnerable to conflagrations. Although a full survey of urban conflagrations in Europe is still lacking, it is clear from numerous scattered references that fires occurred frequently and often caused great damage.[34] Since towns were not very large, the number of casualties was usually low. The fire in London in 1212, in which many hundreds were trapped on London Bridge by a blaze that had jumped the river and engulfed both ends of the bridge, was an exception.[35]

Another similarity with the ancient world was that once a conflagration had broken out, it was practically impossible to stop it, except by removing the buildings in its route of advance. This made prevention and, in the event of an emergency, quick action all the more imperative. From early on, the city authorities therefore issued decrees aiming to reduce the fire risks of buildings, to compel citizens to be careful with fire and to instruct them on how to act properly if a fire did break out.

With regard to building codes, the most obvious measure was to minimize the use of wood and thatch and other combustible materials. The enactment was far from easy, though. Originally, the houses in most European cities were built of wood, or a combination of wood and loam or hardened clay. These materials were preferred over the far more fire-resistant stone and brick, for the simple reason that they were easier to obtain and cheaper. As the French historian Fernand Braudel observed, even Paris was not always a stone city: 'to turn it into one was an immense labour, starting in the fifteenth century', and requiring an enormous workforce, ranging from quarry-workers to masons and plasterers.[36]

Almost everywhere, as long as there was sufficient woodland nearby or upstream, wood was the cheapest building material. It is small wonder, then, that city governments were very wary of making it compulsory to build in stone or brick. At first they promoted the use of stone only by means of subsidies, but they were careful to avoid the prohibition of other materials. It is not difficult to see why they acted so hesitantly. When disaster struck and a town was largely reduced to ashes, it had to be rebuilt in the shortest possible time. Many citizens were ruined and so lacked the means to have their houses rebuilt in stone or brick. In those circumstances, it would have been unrealistic for the city authorities to make building in stone or brick compulsory – unless they could provide wholesale subsidy, which, in the event, they were unlikely to be able to afford. Understandably, therefore, they showed great reluctance to forbid the use of wood and thatch – a reluctance that was reinforced by pressures from the guilds of carpenters and thatchers.

In the end, resistance was overcome, and each town underwent a process of 'stonification' or 'brickification'. Until then, large fires regularly caused enormous damage, without any insurance to compensate for losses. Yet these recurrent disasters were not, in themselves, sufficient cause for the trend towards building in stone to become dominant. The decisive push in this direction seems to have come from the slowly increasing prosperity ('intensive growth') which provided both city governors and private persons with greater resources to spend on building. Thus in the Dutch city of Deventer, for which the process has been well documented, rich individuals were the first to take the initiative and invest capital in houses built of brick.[37] Apparently, as a result of their example, living in brick houses acquired a status value that prompted emulation. Once the voluntary movement towards 'brickification' gained momentum, it became much easier for the authorities to issue general rules requiring that exterior walls and roofs be constructed only from fire-resistant materials. Instead of subsidies for the use of stone or brick, fees were now imposed upon those who continued to build in wood and thatch. What had started as a privilege for the rich – to live in a stone or brick house – eventually became a legal duty for every city dweller.

In addition to the building regulations, the city authorities also issued decrees specifically related to the use of fire. Like the building regulations, these decrees were highly similar from town to town, and many of them closely resembled those already promulgated in ancient Mesopotamia and ancient Rome. Some statutes dealt specifically with crafts in which fire and easily flammable substances were used. Such crafts were permitted only at certain sites, usually on the outskirts of the city, and under particular provisions and restrictions. Thus it was forbidden to use candles in any activities involving oil or flax. When we try to imagine how dark many workshops must have been, we can easily understand how tempting it was to violate such rules, and how difficult they were to enforce. A generally prevailing ordinance for the citizenry at large was the decree to cover all open fires during the night – *couvre feu* in French, and anglicized after 1066 to 'curfew'.

Over the centuries, city governments tended to reissue, with monotonous regularity, the same decrees concerning caution with fire. Apparently, in the name of collective security, they made demands with which many individual citizens did not feel prompted to comply. The long series of statutes seems to consist of countless salvoes in a slow, uphill civilizing campaign.

The same could be said of the third type of prescripts aimed at fire-prevention, those dealing with people's conduct in the event of an outbreak of fire. These prescriptions also clearly put the collective interest above that of the individual. Thus at the first signs of a fire people were not allowed on any account to bring their own belongings to safety; it was even forbidden to start extinguishing the fire at once. Instead, the first thing to do was to rush outside and call the alarm. Any violation of these rules was punished with high fines.[38] Thus the city authorities tried, by external pressure, to suppress the first inclinations of private citizens to let their immediate self-interest prevail over the interests of their neighbours and of the city at large.

Those inclinations were undoubtedly strong. In 1514 a large fire raged in Venice, consuming almost all the shops on the Rialto before the workers from the Arsenal were called in to protect government buildings. Meanwhile, as the American historian Frederic Lane has noted, 'owners of shops, taverns and palaces worked frantically to carry their possessions out of the path of the flames; no one, says the contemporary Marino Sanuto in his detailed description of the catastrophe, gave thought to extinguishing the conflagration'.[39]

The city republic of Venice was exceptional in having no civic fire brigade. In most European towns, taking part in the fighting of fires was regarded as a civil duty. As soon as the tower guards spotted signs of a fire, the bells were rung and the citizens had to hasten to the site. The various guilds each had its own task. Masons and carpenters had to tear down with hooks and pickaxes the walls of the burning houses, so that members of other guilds could reach the blaze with their buckets of water. Quite frequently the fire was too big to approach. Efforts were then concentrated on keeping it within bounds by covering the roofs of adjoining buildings with wet sailcloth and blankets. If this was of no avail, the only remaining measure was to tear down the premises to which the blaze might spread next and to remove all burnable material from its reach. Still, no matter how defective the technical means, something was done in most cities to involve the citizenry in collective efforts to contain and extinguish fires.

I have the impression, from a few scattered references, that it was often difficult to discipline the workers who were recruited to fight a fire, as well as the spectators who flocked to the scene and were either commandeered to join in the handing on of buckets or made to keep their distance. A large fire was sure to attract crowds and to generate tensions. In the turmoil thieves would try to seize their chances. House-owners would protest against the evacuation and demolishment of their property. As in the days of Hammurabi, maintaining public order during a blaze was one of the major concerns of the authorities.[40]

In all these respects, for many centuries fire-prevention in Western Europe did not substantially deviate from the general pattern prevailing in pre-industrial societies. Not until the late seventeenth century were some important innovations in technical equipment introduced, such as the rollable fire-hose, invented by the Dutch engineer Jan van der Heyden. If, before that time, there was anything distinctive about the way Western European cities armed themselves against conflagrations, it was, on the one hand, in the greater effort here than elsewhere on the part of city governments to force their citizens to take preventive measures and to involve them in actual fire-fighting, and, on the other hand, in the gradual process of 'stonification' or 'brickification', which, while receiving stimuli from the authorities, was propelled by its own momentum of rising prosperity and continuing status competition.

Fires

In contrast with the ancient world, for which we have some records of official measures regarding fires and fire-prevention but no detailed accounts of what actually happened during a fire, there are many accurate descriptions of city fires in pre-industrial Europe. Thus, while practically all written evidence about the great fire of Rome in AD 64 is limited to a brief passage in Tacitus' *Annals*, the great fire of London in 1666 has been so meticulously described by Samuel Pepys and other witnesses that we are able to follow its course literally from day to day.[41] The London fire was, of course, exceptional – if only because, during the decades that preceded it, London had rapidly grown to become by far the largest city in England. We also have numerous accurate reports, however, about more typical fires in smaller towns.

Thus the American historian Shelby McCloy quotes some contemporary sources which relate how, in December 1720, two-thirds of the French city of Rennes were gutted. The fire, which went on for almost six days, was said to have started in the home of a drunken carpenter. Fanned by a fierce wind, it spread rapidly through the narrow streets lined by wooden houses with protruding upper storeys. The inhabitants tried in vain to cart their belongings away. The city had but two pumps for fighting fires, and the conduits did not function well. The regiment of Auvergne, in winter quarters at Rennes, received orders to help fight the fire and promote order; instead, the soldiers turned to incendiarism and pillage. Working-class men followed the soldiers in sacking the city, or they demanded extravagant prices for their assistance. The command to destroy a dozen houses in the path of the fire seems only to have added to the chaos.[42]

The wooden houses, the narrow streets, the poor equipment, the deployment of soldiers, the demolition of buildings, the incendiarism, the pillaging and the general disorder – all those elements were characteristic of conflagrations in pre-industrial cities. Technologically, little had changed since Roman times. There was, however, one important difference, concerning the organization and the power relations it implied. In Rome, and in Roman provincial towns such as Nicomedia, fire-fighting was left to a semi-military force that stood directly under Imperial command. The citizens themselves were not allowed to organize a fire brigade (see p. 117). A city in Western Europe, by contrast, would have had its own fire brigade, composed of resident citizens and under the supervision of local officials. In Rennes the citizens even went so far as to disarm the soldiers of the garrison and keep them under guard until the fire was out. The officials apparently had the power to do this, and with impunity at that – something that would have been hard to imagine in the Roman Empire, or in the military-agrarian empires of Asia.

The technical equipment of the fire brigade in Rennes in 1720 was still old-fashioned. Almost half a century before, Jan van der Heyden had patented the rollable fire-hose, which made it possible, for the first time, to reach the seat of a blaze with water. Van der Heyden's invention was not the only one; the second half of the seventeenth century saw a general rise in the standard of fire-fighting appliances. But, as van der Heyden himself knew all too well, material improvements were not enough. He complained bitterly about the 'sluggishness and recalcitrance of the guilds folk', and he tried to reform the organization of the Amsterdam fire brigade, of which he was the commander, so as to create greater discipline and a more professional attitude.[43]

After the great fire of 1720, Rennes also modernized its fire brigade. The town thus followed a general European pattern, in which city governments became increasingly strict about maintaining building regulations and, at the same time, more concerned that the functioning of their fire brigades was up to contemporary standards.

All this contributed to a slow but steady decline in the number of urban conflagrations. Until now, as the economic historians L. E. Frost and E. L. Jones observe, such conflagrations have been treated invariably as 'local events, in an episodic and descriptive fashion'.[44] Yet a systematic compilation by Jones and others of the local data on English urban fire disasters between 1500 and 1900 reveals a clear overall trend. During the first 200 years, the frequency and size of conflagrations in English cities more or less reflected the pattern of urban growth. After 1700, however, the conflagrations gradually diminished, while the cities continued to grow.[45]

Perhaps England, along with the Netherlands, led in this development, but sooner or later all Western European countries joined it. The Great Fire of London in 1666, which destroyed over 13,000 houses, broke all previous records. After that, at least in peacetime, urban conflagrations tended unmistakably to diminish in both frequency and size. In this respect Western Europe differed from the large military-agrarian empires in Asia and eastern Europe. There, even in the capital cities, the great majority of houses continued to be built of wood or wood and clay, and, consequently, great fires continued to occur regularly throughout the eighteenth and nineteenth centuries. The resulting destruction of capital, as Jones suggests, could hardly fail to curb economic growth.[46]

Compensation and Insurance

Another important innovation which emerged in Western European cities was fire insurance. The first clear initiatives were taken in the sixteenth century, in seaports in the Low Countries and Germany. Till then, the only formal rules regarding compensation for damage or loss by fire were rules of retribution, such as those contained in the Code of Hammurabi. Clearly, they were insufficient to deal with any larger calamity. In such an event, the many victims could hope only for relief through charity.[47] The first agencies for them to turn to were local organizations such as guilds, churches and city governments. With the growth of larger state organizations, it became customary, after a big fire disaster, to request aid from the national government and from nationwide church organizations. National governments might grant aid by lump sum and by promises of tax exemption for the near future. All such decisions were made *ad hoc*, as acts of charity.[48]

The insurance schemes were different. They seem to have been inspired by the example of ship-owners who had developed systems of mutual insurance for their ships and cargo – not on charitable but on commercial principles. The first contracts of fire insurance were drawn up along the same lines. Business competitors committed themselves by contract to put money into a joint fund out of which each contributor would receive compensation in the event of property loss through fire.[49]

Apparently fire insurance became a profitable business after the London fire of 1666. By 1720 there were six fire-insurance companies established in London, bearing such confidence-inspiring names as the Friendly Society for Securing Houses from Loss by Fire and the Amicable Contributorship, commonly known as the Hand-in-Hand Society. In the eighteenth century, starting from London, commercial fire insurance slowly spread through Great Britain.[50]

Meanwhile, on the Continent some attempts were made to turn fire insurance into a government operation. Thus in 1705 Frederick I of Brandenburg set up a project of fire insurance for buildings that was compulsory for all proprietors, at a low standard rate. The scheme met with strong resistance from the start, and when, after a major disaster in 1708, the treasury was unable to pay indemnity, the king had to abolish it. However, in 1718, under his successor Frederick William, the institution of a mutual fire-indemnity society with compulsory membership for the city of Berlin did prove viable. It turned out to be one of 'the forerunners of a continuous development of public, mutual and joint-stock companies, insuring both real and personal property, in the following three centuries'.[51]

We do not have a single term to cover the rapid growth of insurance in the modern world after its slow start in pre-industrial Europe. Fire insurance was just one variety among many. In spite of the initiatives by the kings of Brandenburg, it continued to be arranged mainly on a voluntary basis. Entering into a fire-insurance contract was, for both parties, a strategy of behaviour informed by the calculation of long-term risks. For the individuals concerned, it would always be a decision based upon 'rational choice'. However, this 'rationality' could function only if a great many people acted in accordance with it. The changes in conduct and attitude that were involved in the growth of fire insurance were, in common with many other aspects of the European civilizing process, the manifestation of greater interdependence between increasingly large numbers of individuals.[52]

Air Pollution and Fuel Supply

In addition to the risk of conflagrations, the use of fire in the cities of pre-industrial Europe also caused other problems. Of these, the supply of fuel of course came first, but in popular attention air pollution often took precedence. Smoke could indeed be a great nuisance. As long as ventilation was poor, the problem would be felt indoors, directly around the hearth. The introduction and gradual improvement of chimneys did a great deal to reduce this domestic inconvenience. As a result, almost inevitably, people tended to burn more fires and thus to consume more fuel and to produce more smoke. In the long run, this led to a shortage of wood and its derivative, charcoal, and to the use of inferior fuels, causing serious problems of air pollution in many cities.

Traditionally, wood was by far the most important fuel, but it was a bulky product, unwieldy to transport long distances overland. The possibility of obtaining it without unduly high costs, either overland from nearby or otherwise by water, was one of the conditions determining the location of cities, and of urban industries.

In several parts of Europe fuel shortages became acute as early as the thirteenth century. One of the first cities to suffer from 'the gradual elimination of forests at the urban perimeter' was London.[53] As the population doubled from 20,000 in 1200 to 40,000 in 1340, wood as a fuel was increasingly replaced by a low-grade coal called 'sea coal' that was mined near Newcastle and brought to London by ship. Sea coal emitted a foul-smelling smoke, penetrating everywhere, and left behind an omnipresent deposit of black soot. Its fumes were said to have driven Queen Eleanor out of the city at the time of the feast of Saint Michael in 1257. After the Black Death in the mid-fourteenth century, the population declined, and so did the consumption of sea coal. But the relief was only temporary, for by the second half of the sixteenth century London entered a period of great prosperity and growth, and the consumption of sea coal became higher than ever before.[54]

Another alternative fuel was peat. Many parts of Europe were, and still are, covered with peat bogs from which this fossil fuel could easily be obtained. Even after intensive drying, however, peat was still more voluminous than wood, and therefore very costly to carry overland. In this respect the Low Countries had the good fortune of possessing waterways over which peat could be transported cheaply. According to the Dutch agronomist J. W. de Zeeuw, it was by virtue of the availability of this alternative fuel that the Dutch urban economy and culture could flourish as they did in the seventeenth century.[55]

In England, less favoured with inland waterways, other solutions were sought for the fuel problem. Increasingly, coal was used – at first, as a substitute for wood and its derivative, charcoal, but later as a fuel in its own right. As the British historian A. E. Wrigley points out, the increasing use of coal involved a transition from an almost complete dependence on organic energy *flows* to a progressive dependence on fossil *stocks* of energy. By the end of the eighteenth century, in order to supply England with the equivalent in wood of its annual coal consumption, the country would have needed many more millions of acres of woodland than it actually had.[56]

FIRE IN THE COUNTRY

Deforestation

Throughout the entire period from 850 to 1850, the great majority of people continued to find their living in the country. It may seem odd to focus upon them only after having discussed the use of fire in religion, in war and in cities, but then, of course, a great deal of the discussion was also about them, if only implicitly, for they were the ones who formed the great majority of believers, on whose lands battles were fought and who supplied the urban population with food and fuel.

By and large, the rural world showed a slower pace of change than the towns. Yet it was also undergoing transformations. One of the most momentous of those was the gradual expansion of arable and pasture at the expense of forests. After the collapse of the western Roman Empire, many cleared woodlands had been abandoned and were overgrown again with trees, so that, by the early Middle Ages, most of Western Europe – unlike large parts of China and northern Africa – was covered with dense forest. By the end of the nineteenth century, however, only a few of these forests remained. As causes of this process of deforestation, the needs both for wood and for open fields and pastures supplemented each other. Part of the wood was used as timber, for buildings and ships, but probably by far the greater part was consumed as fuel.

The first five centuries after 850 were marked by virtually uninterrupted extensive and intensive growth, and deforestation proceeded accordingly. The population decline following the Black Death, around 1350, made for a temporary setback, but in the fifteenth century population began to grow again and, concomitantly, land use became more intensive, and deforestation resumed its pace. By the second half of the seventeenth century, some densely populated regions, such as the Netherlands and England, were largely depleted of woodland. In England only parts of the protected royal forests escaped felling; in the western part of the Netherlands not a single forest survived.[57] As pointed out above, the two countries found very different alternatives for wood as a fuel.

Since the practical value of wood was high, we have little evidence of slash and burn practices. In *The Agrarian History of Western Europe, AD 500–1850* by the Dutch historian B. H. Slicher van Bath, they are hardly mentioned. People would not wantonly burn trees; even if they were primarily interested in clearing land, they would burn away only roots, branches and undergrowth, as wood was far too precious in the agrarian and urban economy.[58] Old forms of swidden agriculture survived only along the slowly receding borders of the cultivated European mainland, in the 'frontier areas' of Russia and Finland. Within those fringes, rural and urban Europe alike had become 'fire-protected zones'.

Fire Use and Fire Hazards

In the rural world of pre-industrial Europe, fuel was rarely used in great quantities. In the households, the same hearth that was used for cooking also served as the main source of heat and light. During the winter, the people themselves and their animals were additional sources of heat; if extra light was needed, a candle or an oil lamp had to suffice. Tolerance for cold was probably much greater than what we are used to nowadays – even the royal palace at Versailles was so poorly heated that, during a dinner in the luxurious Hall of Mirrors in the winter of 1695–6, the wine and the water froze on the table.[59]

Local craftsmen such as blacksmiths would use far more fuel than the average household. Yet the only really large consumers of fuel were specialized industries such as iron mines, lime kilns and brick factories, which formed outposts of the urban economy rather than integral elements in the agrarian economy. Charcoal-burning, a form of fuel production which required considerable quantities of fuel to begin with, was usually done in the forests, and affected the rural world mainly as a source of employment.

Fire hazard was a perennial worry in the rural world of pre-industrial Europe. Although in the country houses did not stand as closely together as in the cities, they constituted great fire risks because of the flammable material of which they were built. At a time when houses in the cities had long ceased to be fitted with wooden fronts and thatched roofs, wood and thatch were still the favourite building materials in many parts of rural Europe. As late as 1854 a 'thatched-roof rebellion' broke out in the Angevin countryside in north-western France, when the prefect decreed that all thatch had to be replaced by slate and tile. Apparently the walls and timberwork of

Fire and Civilization

most peasants' huts were too weak for heavier covering and would have had to be strengthened, if not totally rebuilt:

Many peasants too poor to bear the cost of a new roof, let alone the cost of rebuilding their houses in stone rather than clay and wood, resisted the order and were evicted. They marched on Angers, several thousand strong, and the army had to intervene and disperse them. The prefect was replaced, the decree rescinded, and thatched roofs allowed to disappear more gradually. The relatively prosperous farmers who carried insurance were finally forced into compliance under a concerted attack by insurance companies in the 1860s and after.[60]

Fires might have many causes. One against which people were completely powerless was lightning. All over Europe, farmsteads and sheds were decorated with magical spells and pictures intended to avert this danger. In his book on fire folklore in Germany, Herbert Freudenthal has collected an astounding variety of often contradictory beliefs about objects and incantations that might keep lightning at bay. It is a far cry from these desperate charms and superstitions to the clear and simple certainty with which Benjamin Franklin announced, in 1752, that he had discovered 'how to secure houses, &c. from lightning'.

It has pleased God in his Goodness to Mankind, at length to discover to them the Means of securing their Habitations and other Buildings from Mischief by Thunder and Lightning. The Method is this: Provide a small Iron Rod (it may be made of the Rod-iron used by the Nailers) but of such a Length, that one End being three or four Feet in the moist Ground, the other may be six or eight Feet above the highest Part of the Building. To the upper End of the Rod fasten about a foot of Brass Wire, the Size of a common Knitting-needle, sharpened to a fine Point; the Rod may be secured to the House by a few small Staples. If the House or Barn be long, there may be a Rod and Point at each End, and a middling Wire along the Ridge from one to the other. A House thus furnished will not be damaged by Lightning, it being attracted by the Points, and passing thro the Metal into the Ground without hurting any Thing. Vessels also, having a sharp pointed Rod fix'd on the Top of their Masts, with a Wire from the Foot of the Rod reaching down, round one of the Shrouds, to the Water, will not be hurt by Lightning.[61]

Franklin's invention, made in Philadelphia, rapidly reached Europe and then gradually penetrated the rural world. In spite of the proud advertisement that 'a House thus furnished will not be damaged by Lightning', several generations passed before most people were convinced of its efficacy. Among the resistances to be overcome were not only objections from theologians but also a widespread scepticism towards novelties originating from cities. In this case such scepticism found support in the common-sensical thought that it would be foolish to put a contraption upon one's roof that was designed with the very idea of attracting lightning!

Another frequent cause of fire in the country was hay heating – a chemical reaction that could result in spontaneous ignition, unless the hay was regularly inspected and turned over. In the event of a fire started by hay heating, the causes lay in a combination of natural processes and human failure to intervene in time. Most fires, however, were the direct consequence of the use of domesticated fire. Quite often they could be attributed to carelessness, and in many cases the blame was put on older women or children.[62] Deliberate incendiarism also occurred frequently, however. There were times when people in the country felt themselves as helpless against arson as they were against lightning.

Incendiarism

In settled agrarian societies the deliberate destruction by fire of other people's property has always ranked as a serious offence, second only to murder.[63] Like murderers, arsonists were generally punished by death. In English criminal law, hardly any attempt was made to differentiate according to the gravity of the offence: capital punishment applied even to the burning of a haystack or a barn. Nor was the variety of possible motives taken into account, although, as the legal historian Leon Radzinowicz noted, these motives might 'equally well be a desire to attain material gains, vengeance, or even – particularly among agricultural wage-earners – the expression of social unrest'; in addition, arson might 'be due to drunkenness, or be the symptom of a morbid personality'.[64]

The variety of individual motives should not make us disregard the social context. As the French sociologist Emile Durkheim demonstrated, even the incidence of such a pre-eminently individual deed as suicide clearly varied with social circumstances; it made sense, therefore, to regard the act of suicide as a desperate last solution sought by people for problems resulting from the social situation in which they found themselves.[65] The same might be said of arson.

Like suicide, arson is a phenomenon the actual occurrence of which may often be difficult to ascertain. Especially in modern society, in which defrauding insurance companies has become a common motive for arson, arsonists will do all they can to erase every trace of their action. Yet the incidence of arson committed by people in order to collect insurance money is a function of particular conditions in contemporary society; as such, it clearly belongs to the category of what Durkheim would call 'social facts'. Similarly, in the rural world of pre-industrial Europe, there were forms of socially induced incendiarism.

The known details about this rural incendiarism are still fragmentary. Nevertheless, we can sketch an outline of the configuration in which it generally occurred. This consisted of three parties: first, farmers living in isolated farmsteads, bound to their property; second, a rudimentary police force; and, third, potential incendiarists, ranging from anonymous vagrants to disgruntled servants and labourers.

Many acts of arson were committed by individuals acting on their own, often motivated by a sense of grievance against their present or former employer. Sometimes, however, groups of people collectively threatened arson. Such threats might emanate from bands of tramps (or, in some cases, local people) extorting money, or from groups demanding some sort of social reform. Vagrants were able to voice their threats directly; they could use certain codes such as the warning to a farmer that, on the next morning, he might be 'woken up by the red cock'. Local people had to hide their identity; they communicated by written message, scrawled in clumsy handwriting, demanding that money be deposited at a designated site and saying that if addressees failed to oblige, they would be 'rewarded with ashes'.[66]

The incendiarists conducted a sort of mini-war beyond the reach of the state's central monopoly of organized violence. Some bands espoused idealistic motives. Thus, in the middle of the sixteenth century, groups associated with the Anabaptists terrorized the countryside in the eastern Netherlands. After they had burned down a number of farmsteads, they were increasingly considered a menace to the established order. Urban and regional authorities united to persecute them. Faced with this superior power, the movement lost part of its following; the remaining core members were hunted down and executed.[67]

This was to remain a common pattern in the rural regions of pre-industrial Europe. Vulnerability to fire, and therefore also to arson, was part of the general precariousness of rural life. The risk of incendiarism was always present; it became particularly acute during periods of mass vagrancy or when central authority was at its weakest. In areas far removed from central state control, bands of arsonists sometimes managed to hold out for many years, but they were never able to stand up against a well-organized military force.

As long as a region was plagued by incendiarists, this was a serious impediment to saving and investment, and therefore to 'intensive growth'. People could enjoy a steady accumulation of property only if they knew themselves to be reasonably well safeguarded against arson. Such safeguards might be attained either through self-protection or through protection by force of law, or because – for whatever reason – banditry did not occur. The major form of self-protection comprised solid stone walls with well-locked gates, but this was a luxury which in itself required a level of wealth far above the reach of most peasants. In the distant regions of pre-industrial Europe, the force of law could only sporadically manifest itself. And the general level of wealth and collective welfare was such that again and again new waves of vagabonds emerged to roam the countryside – in their midst were likely to be at least some who themselves had been the victims of a fire and had lost all their possessions.[68]

No comprehensive survey has yet been made of the waves of collective incendiarism that affected the rural regions of Europe in successive centuries. For other parts of the world the picture is even less complete, but it seems unlikely that the rural population there would have found the means to avoid this scourge. As in the Roman Empire, the threat could come both from 'below' and from 'above', from tax collectors using arson to lend force to their impositions. During riots feelings of popular indignation, revenge and the desire to destroy tax registers or incriminatory evidence all combined to make the houses of the tax collectors, in turn, a favourite target for arson – along with court houses and other government buildings.

Since incendiarism was very much feared, contemporaries may sometimes have exaggerated its scale. But given how easy it was to set fire to a farmstead, the actual record of arson may still be regarded as relatively low; it could have been much higher. In mutual conflicts between farmers, recourse to this *ultima ratio* was taken only rarely – although the French sociologist Gabriel Tarde did write in 1895 that 'in the country, setting fire to the barn of one's enemy ... is the means of revenge employed most frequently and with most impunity'.[69] And we do indeed hear of occasional epidemics of incendiarism, such as the one that ravaged a village in the Gévaudan in the 1840s, when one house after another was gutted in a chain of revenge and counter-revenge.[70] Normally, however, social relations were in this respect characterized not just by mutually expected but also by mutually exercised self-restraint.

With the penetration of central state control, the menace of rural incendiarism during peacetime was gradually suppressed. In the mid-nineteenth century, several rural regions in southern England were the scene of great poverty and social upheaval, accompanied by arson. According to the British historian David Jones, 'at a conservative estimate, the worst years saw a minimum of 1,000 fires, and some of them were very large indeed'. The 1860s marked a turning point: incendiarism declined and was replaced by 'more open and peaceful forms of protest, namely meetings, petitions, strikes and early trade unionism'.[71] In France, 'aggressive pauperism' persisted even longer, in a less clearly articulated political guise. As late as the end of the nineteenth century, remote rural districts continued to be visited, in bad years, by bands of vagrants who threatened arson.[72] The twentieth century saw such bands virtually disappear, thanks to the strengthening of the forces of the police and the law, in combination with a general rise in the level of welfare provisions. Thus intensive growth created the conditions under which, at least for the time being, a chronic threat to intensive growth itself came to an end.

FIRE IN TECHNOLOGY AND SCIENCE

Every section of society was affected sooner or later by a series of advances in the control of fire – in the capacity to understand processes of combustion and to utilize them without being injured by their destructive force. This development as a whole, although punctuated by some major discoveries and inventions, was gradual. The set of transitions that we can distinguish with the benefit of hindsight, from alchemy to chemistry, from craft to industry, from magic to science, was in fact a multitude of small innovations in glass-making, pottery, metallurgy, lime-burning, the distillation of alcohol and so on. They cannot be reduced to single events that happened at a particular place and time.

In the Middle Ages, metallurgy and alchemy were probably the most important catalysts in the development of the control of fire. Both areas were highly competitive and were accordingly characterized by peculiar mixtures of secrecy and openness, profit-seeking and pure curiosity. The fire masters, working on the borders of craft, black art and science, recognized a certain hierarchy of prestige, but they were not subject to any official authority.

Thanks to this rather open social structure, a principle of 'cumulative causation' could operate in the development of the control of fire. Advances in various areas stimulated each other. Improvements in the construction of furnaces made it possible to produce higher and more even temperatures. This led to the development of new metal and glass instruments which enabled the investigators to conduct still more experiments with fire. Among the many 'radiations' of enhanced control of fire were improvements in book printing made possible by the casting of movable type in an alloy of lead, tin and antimony. In turn, books contributed to the propagation of knowledge, including knowledge about fire, as contained, for example, in Georgius Agricola's impressive treatise on metallurgy and mining (1556).[73]

The workshops of artisans continued to be the major setting in which fire techniques were learned and employed. These techniques were acquired through apprenticeship and personal experience. In order to appreciate how important personal experience was, we have to realize that there were, for example, no instruments for objectively measuring temperatures. Authors of manuals were able to give only rough indications in vague categories, varying from 'seizable by hand', to 'just touchable', to higher degrees of heat for which adequate terms were almost wholly lacking. Only at the beginning of the eighteenth century did such men as Gabriel Fahrenheit, Anders Celsius and R. A. F. de Réaumur succeed in devising instruments and scales based upon the expansion coefficients of alcohol and mercury. From then on it became possible to perform quantitative measurement at least for the range of temperatures between the freezing point and the boiling point of water. For more than two centuries, however, different nations continued to use different scales.[74]

Besides thermometers, the eighteenth century also brought increasingly accurate weighing instruments. These could be used to acquire a better insight into the nature of combustion processes. This spelled the end for the ancient theory that fire was one of the four elements, along with air, water and earth. Until well into the seventeenth century, this idea was held to be self-evident and formed the context for practically every alchemical or chemical experiment. Its import was still evident as late as 1720, when the Dutch scientist Herman Boerhaave declared that in order to resolve the central enigma of the universe one had first to resolve the enigma of fire: 'If you make a mistake in your exposition of the Nature of Fire, your error will spread to all the branches of physics, and this is because, in all natural production, Fire . . . is always the chief agent.'[75]

In the course of the eighteenth century, thinking about fire was thoroughly transformed. A change took place which had been adumbrated as early as 1611 in a prophetic line by the poet John Donne:

'The Element of fire is quite put out.'[76] The theory of the four elements received its first scientific blow from the so-called phlogiston theory. According to this, combustion (like rust) was a process in which an invisible substance (phlogiston) was released; in the case of fire, the process was accompanied by heat and light. Although the new theory helped to clarify many riddles, it left one problem unresolved: how was it to be explained that in the process of combustion (in other words, in the process of releasing phlogiston), many substances seemed to gain rather than lose weight? For a long time, scientists tended to attribute the anomaly to flaws in the measuring and weighing equipment.[77] Then, in 1777, Lavoisier demonstrated that combustion (as, indeed, is true with rust) consisted of a process of combination with oxygen. Once this discovery was made, and the conclusions were corroborated by numerous experiments, the theory of the four elements had to be given up in the sciences, to be replaced by a far more complicated system leaving room for many scores of elements. Soon afterwards, the concept of fire no longer appeared in the textbooks; the new specialism, thermodynamics, recognized only heat and energy.

As an empirical phenomenon of great significance, fire did not vanish so quickly. At the same time as the concept of fire was banned from the physical sciences, huge furnaces were built, with chimney stacks dominating the skyline. On an increasingly large scale, fire was used to propel engines and vehicles. The beginning of the industrial era, just like the beginning of the agrarian era, was marked by a substantial intensification in the use of fire.

8. FIRE IN THE INDUSTRIAL AGE

===

INDUSTRIALIZATION AS A DOMINANT TREND

After the original domestication of fire and the rise of agriculture and animal husbandry, industrialization was the third major ecological transformation set in motion by humans. For some time, it was commonly referred to as the Industrial Revolution – a term that evokes the image of a radical break with the past occurring in England between 1780 and 1850, as the British counterpart to the French Revolution. This view has now generally been abandoned. Most historians today agree that the origins of industrialization go back further than 1780, and also that whatever 'revolutionary' impact it had did not make itself felt on a large scale until after 1850.[1]

Industrialization presents itself as a 'congeries of changes' that seems more complex and less easy to define than the two preceding transformations.[2] Yet it is possible to find a common denominator in this congeries of changes. Of crucial significance was, once again, the incorporation into human society of natural forces that were previously outside the human realm. These natural sources were first of all, as E. A. Wrigley points out, large, untapped sources of energy stored up in fossil fuels.[3] Tapping increasingly more sources of coal, and then of oil and gas, enabled people to exploit, on a much larger scale than ever before, a great variety of other mineral sources, from iron to plutonium.

Like the preceding transformations, industrialization did not appear out of the blue. Clearly, industry in the sense of the manufacturing of objects had been practised by humans since the Palaeolithic. Metals were being processed as early as several thousand years ago — as the very names Bronze Age and Iron Age are intended to convey. Nor was coal entirely unknown as a fuel. Until recently, however, all industrial activities were embedded in a predominantly agrarian structure; they did not alter the overall agrarian character of society.

The transition to a predominantly industrial world was gradual, and it would be futile to try to locate precisely when and where it began. Yet the convention of giving pride of place to England is not altogether wrong. There, in the eighteenth and nineteenth centuries, many rivulets of 'proto-industrialization' merged into a swelling and irreversible stream. In due time, the consequences were felt all over the world and in every sphere of life. They included an unprecedentedly high absolute rate of both extensive and intensive growth, and a continuation, at a faster pace than ever before, of such related trends as the increasing concentration, specialization and organization of the human population.

Extensive growth has been momentous over the past 200 years. In 'the first industrial nation', Great Britain, the population increase throughout the nineteenth century was in the order of 11.1 to 16.9 per cent each decade — a figure that would have been considerably higher still had there not been a steady stream of emigration to the United States and the colonies.[4] For the world at large, human population has increased from an estimated 900 million in 1800 and 1,600 million around 1900 to more than 5,000 million by the end of the twentieth century.[5]

The figures indicating intensive growth are possibly even more impressive. Thus the British geographer and ecologist I. G. Simmons cites some figures indicating the rise in energy — expressed in the standard unit of MJ or megajoules — available to people at various stages of socio-cultural development. Before the domestication of fire, the average was about 10 MJ per day, which is the somatic energy needed to keep a single human being alive. Then, as various extra-somatic sources of energy were added, this average increased to about 100 MJ in the most advanced agrarian societies, while today, the per capita commercial energy consumption in the USA is close to 1,000 MJ per day. As Simmons concludes, there has been a rise of two orders of magnitude between the Palaeolithic and the present. The total quantity of energy now being converted is unprecedented, and the rate of growth in recent decades has been twice that of the human population, although some slowing up of both rates has been characteristic of the last few years.[6]

Wherever industrialization first made itself felt, it had the effect, initially, of further increasing those differences in power and conduct, both among and within societies, that had arisen in the age of agrarianization. The gap widened between the leaders in the process of industrialization and those societies which lagged behind, so that in the second half of the twentieth century it became generally customary to distinguish between First, Second and Third World countries – just as, in the middle of the nineteenth century, England was said to consist of 'two nations'.

At the same time, however, industrialization, like the domestication of fire, has unleashed forces which have led to a decrease of differences in power and conduct. The industrial regime under which people everywhere have come to live exerts, just like the fire regime, certain uniform pressures evoking essentially similar responses. If the complaints often voiced about the standardizing and levelling tendencies of modern life have any empirical ground, this is likely to be it.

THE AGE OF THE STEAM ENGINE AND THE SAFETY MATCH

There is a plausible and direct association between the image of the Industrial Revolution and the steam engine. Engravings from the first half of the nineteenth century show the British industrial landscape dominated by factory stacks emitting huge clouds of smoke. At night the ruddy glow of furnaces and steam engines enveloped the entire surroundings. Travellers such as Alexis de Tocqueville or Charles Dickens, who came from regions that did not yet so clearly bear the imprint of industrialization, were deeply impressed.[7] In the same vein the Amsterdam merchant H. P. G. Quack described in his memoirs the unforgettable view of Liège when, around 1860, as a young man he approached the city by train at dusk:

On all sides the fires of the foundries are shining, on all sides you hear the dull humming of the engines, the steam is whistling, the dark smoke billows in black plumes along the track, or the vapour rises up in a white thin column. The steam boats glide by you on the river; the locomotives thunder past you with their trains.[8]

For more than a century, the awed observations of contemporaries continued to resound in the writings of historians. In their estimation, the steam engine was 'the pivot on which industry swung into the modern age'; 'steam enabled the rapid and universal development of large-scale industry to take place'; they described 'the invention of the steam engine' as 'the central fact in the industrial revolution'.[9] A younger generation of historians, however, have re-examined the evidence and come to the conclusion that the role played by the steam engine has been overestimated.

By the first half of the eighteenth century, there were indeed a few big industrial plants in which goods were produced on a large scale by factories, but without any use of steam. Thus some English silk factories employed hundreds of labourers using mechanical tools driven by water power.[10] The steam engines which were developed in the course of the eighteenth century were used mainly in mines. Their share in the process of production was still very modest. According to an oft-quoted calculation by the economic historian G. N. von Tunzelmann, in the hypothetical case of there having been no steam engines at all, economic growth in England over the entire eighteenth century would have been retarded by less than two months.[11]

The heyday of the steam engine did not come until the second half of the nineteenth century. By that time, factories with water- or wind-driven engines were no longer able to compete with factories equipped with steam power. But even before this period the steam engine probably exerted an indirect influence, since pressure of competition forced the owners of water-mills to enlarge the capacity of their plants to a maximum. A similar effect was seen later in seafaring, when after the introduction of steam ships a completely new type of sailing ship was developed, the clipper, which for several decades was as fast as a steam ship.[12]

Generating steam power demanded larger investments than were required for water or wind power. Wherever and whenever water and wind were available, they were, in principle, free, whereas steam always had to be produced, in operations requiring labour and fuel. But while water and wind were cheaper energy sources than steam, they could be tapped only in limited quantities, and their availability depended entirely on natural conditions which were beyond human control. Moreover, before the invention of generators for producing electricity, there was no way of transporting either water or wind energy. The mills were therefore bound to specific locations; and windmills had the additional drawback of being completely dependent on the weather.

For the utilization of steam power, some time-honoured principles of fire control − fire generates fire, and both fire and fuel are transportable − could be applied. In the long run, this gave entrepreneurs who worked with steam engines an invincible edge over competitors who stuck to wind- or water-mills. They had to reckon far less with natural restrictions in choosing a location for their plants; they could let their engines run with great regularity, twenty-four hours a day, in every season; and, by installing heavier machinery, they were able to increase the productive capacity far beyond the reach of wind- and water-mills.

Other mechanisms which we encountered earlier in the history of the control of fire also applied to the steam engine as well. By using steam power people made themselves less directly dependent on certain natural forces such as streams or wind; but they still remained dependent on 'nature' − in this case, on the presence of supplies of coal. Moreover, as their dependence on forces of nature became more indirect, their dependence on other people increased. The pressures of social interdependence became at once more powerful and more diffuse and elusive.

The industrialists were affected by these pressures from various sides. Even if they considered themselves to be independent entrepreneurs, they had to deal with suppliers, customers, competitors and employees. The part they played in this multipolar social figuration impelled them continuously to make new investments in order to increase efficiency and productivity. The resulting overproduction caused recurrent recessions and bankruptcies. With each crisis, some competitors vanished, leaving a smaller number of increasingly large companies. This unplanned process of the formation of economic monopolies, which was subjected to penetrating analysis by Karl Marx, was later characterized by Norbert Elias as a typical 'elimination contest'. Structurally it resembled earlier elimination contests, such as the political and military struggles fought by feudal lords in the late Middle Ages, out of which the states of modern Europe eventually emerged.[13]

Those industrial capitalists who survived in the economic elimination contests were able to make enormous profits. In stark contrast to their increasing wealth stood the poverty of the workers. While the capitalists as a class were under permanent pressure to install larger and more expensive machinery, most individual members of this class were free to leave the competitive struggle and to pursue a different career or to live as rentiers. Workers had no such options. They became exclusively dependent upon the wages they received for their contribution to the industrial production process. As the steam engine became the pivot around which industrial production revolved, the workers had little choice but to submit to the rhythm it imposed upon them: 'All the machines were geared to the engine, and the entire sequence of production demanded that each worker subordinate his own will to that of the whole working unit.'[14]

New forms of discipline on the factory floor were one effect of industrialization. Another was the general *Verelendung*, the worsening of the conditions of the working class. The term was coined by Karl Marx, but the process as such was visible to any bourgeois visitor who ventured into the factories or the mines or into the slums where the workers lived. The long working hours, the low wages, the poor housing, the permanent threat of unemployment and the absence of any reserves to meet crises combined to make the workers' lives miserable. One did not need to be a socialist to recognize this. Thus

Alexis de Tocqueville noted after his visit to Manchester in 1835:

Look up and all around this place and you will see the huge palaces of industry. You will hear the noise of furnaces, the whistle of steam. These vast structures keep air and light out of the human habitations which they dominate; they envelop them in perpetual fog; here is the slave, there the master; there is the wealth of some, here the poverty of most; there the organized efforts of thousands produce, to the profit of one man, what society has not yet learnt to give . . . Here humanity attains its most complete development and its most brutish; here civilization makes its miracles, and civilized man is turned back almost into a savage.[15]

At the same time as contrasts were increasing within the industrializing countries, the differences in power and behaviour between those countries that were and those that were not yet industrializing grew larger as well. Thanks to their flourishing industries, the European states developed formidable military power to which almost every other society had to submit. The natives of North America and Australia were brought near to total annihilation; many parts of Asia and Africa were subjected to colonial administration, and large parts of the land and the people were put into the service of the production of raw materials for industry.

The military, political and economic superiority of the European nations relied heavily on their advanced industrialization. At the same time, however, there were, in the very processes of industrial production, tendencies towards a diminution of the differences in power and conduct, both among and within nations. One of these tendencies was the trend towards mass production, resulting in the standardization of industrially manufactured commodities. A typical product of this trend was the safety match.

The history of the match as a device for making fire may be said to have begun in the seventeenth century, with the discovery of phosphorus. There is some synchronism, which is not altogether fortuitous, between its development and that of the steam engine. The advances in specialized technology and science, combined with the general rise in prosperity, created the conditions for a deliberate search for new fire-making devices which would be cheaper to produce and less cumbersome to operate than a tinderbox. The spread of the habit of smoking especially created a wider demand for such devices. Towards the end of the eighteenth century, several sorts of match containing phosphorus were invented. In spite of many improvements introduced in the following decades, however, they continued to be unsatisfactory in several respects. They gave off a shower of sparks when lit, and they could easily ignite spontaneously, through an abrupt movement or if they happened to be exposed to the sun; moreover, the heads were made of a material ('yellow phosphorus') that was very poisonous. These disadvantages were first overcome in 1852 with the invention, patented by the Swedish manufacturer Johan Lundström, of the safety match, which would ignite only when rubbed against the surface of the box – a surface containing the relatively harmless 'red phosphorus'.[16]

The sociologist Herbert Spencer is reported to have called the safety match 'the greatest boon to mankind in the nineteenth century'.[17] This may seem an unduly exuberant statement, but the idea behind it was that the safety match made fire available to everyone, all over the world, with a minimal risk of accidents, and at a cost that soon became truly negligible. Matches, as has often been noted, were the only human artefact so cheap that people might ask even a stranger for one. In the second half of the nineteenth century, their use spread rapidly through all sectors of the population in the industrializing nations, and wherever Europeans penetrated, the matches they brought with them were among the very first elements of material culture adopted by the people with whom they came into contact.

We may well wonder why it was that matches became so popular. The immediate answer would be that they made fire available to every person at any time. But then we could ask why people should wish to make fire. It was estimated that by the 1870s the average English person struck eight matches per day.[18] Why did they do this? In the household, they would need fire for cooking, for heating and for lighting their oil and gas lamps. In addition, increasing numbers of men needed fire for smoking. Apparently, the level of technology, prosperity and comfort to which more and more people became accustomed helped to create a market for the match industry.

The industry itself soon developed into a textbook case of

monopoly formation. By the beginning of the twentieth century
world match production was controlled by a few companies, with
mutual trading agreements and interlinked financial interests. For
people everywhere, matches had become an indispensable need. The
manual skill involved in striking a match was very easily acquired;
the most important thing people had to learn was to observe the
necessary caution.

Safety matches were simple things, but they would have been very
difficult for individuals to make for themselves. Not only did most
people lack the knowledge that was needed to manufacture them,
but they would also have been unable to obtain the necessary
materials; nor would they have had the machinery for processing
those materials. Safety matches were just one of a great number of
objects for which people in an industrial society were wholly dependent on other people – people of whose existence they might not
even have been aware.

NEW SOURCES OF ENERGY: MORE DISCRETE
AND DIFFUSE USE OF FIRE

The industrial landscape has changed thoroughly in the course of the
twentieth century. It is no longer dominated by endless rows of
factory chimney stacks. The presence of fire and smoke has become
far less conspicuous.

These changes in the landscape reflect the disappearance of steam
power as the main source of energy in industry. It has been superseded
by other sources such as oil, gas and electricity. But while at first sight
the part played by fire in industrial production appears to have
diminished, a closer look reveals its continuing importance.

In fact, the methods of production in modern industry, and in
agriculture as well, are highly fuel-intensive. Most of the energy
consumed – including most electricity – is derived from the fossil
fuels coal, oil and gas. Combustion processes continue to play a
central role, but they are relegated to special containers so that most
people are not directly confronted with any of the features that can

make the presence of a fire annoying and dangerous. The caprices of the flames are fully controlled. Soot, smoke and fire risks are reduced to a minimum. Furnaces and combustion chambers in which great heat is concentrated remain cool on the outside.

Typical products of modern fuel-intensive industry are motor cars, with engines designed to be propelled by finely tuned and minutely controllable combustion processes. Indeed, the motor car can almost serve as a symbol of the highly complex and differentiated ways in which, at the end of the twentieth century, fire is used. Cars are set in motion by burning fossil fuel. They are made of steel, plastic and glass – materials produced and processed at high temperatures. Yet someone who gets into his or her vehicle and turns on the electrical ignition to start the engine is not likely to be consciously aware of using fire and the products of fire. When driving, people do not perceive the processes of combustion which keep their car going; they do not see the petrol gas burning under the bonnet, nor have most of them even remotely sensed the fire in the factories and power plants without which their cars would never have been produced at all.

A very different example of the same effect is farming. In the beginning of the nineteenth century, when England was already beginning to industrialize, practically all the energy consumed on the farm was produced within the confines of the farm and its fields, in the form of human and animal labour; the open fire that was burning in the hearth was fuelled with wood from the immediate sur-roundings. By the end of the twentieth century the situation has become very different, with practically all the energy used now brought in from outside the farm, in the form of fertilizer, oil and petrol, and electricity.[19]

A major advantage of the new sources of energy, not only over wood but over steam as well, is that they can be applied with greater flexibility. The fuels are easier to transport and to distribute than wood or coal, and the combustion can be regulated more precisely. Given the necessary technical facilities, gas, oil and electricity provide for very even and accurately controllable flows of energy. Electricity has the additional advantage of being totally 'clean' at the place of destination.[20] Almost all the chores and nuisances connected with the use of fire, such as getting rid of smoke, storing fuel and tending the fire, are no longer necessary for the consumers of electricity. Fire hazards have not been eliminated altogether, but they have been greatly reduced. People are able, nowadays, with a few simple actions and a minimum of risk, to avail themselves of large quantities of highly concentrated energy.

This is so in every area of life, be it agriculture, industry, traffic and transportation, domestic work or leisure. Everywhere it is possible to mobilize large quantities of energy with very little physical effort. The result is to make life more comfortable in many respects, enhancing the sense that physical processes can be mastered and, concomitantly, creating the illusion of independence.

An illusion it clearly is. Regardless of whether people can avail themselves of energy in the form of a petrol engine, a battery, or a connection to an electric circuit or a gas main, in each case they are able to do so only because they are part of a complex and far-reaching network of social interdependencies. As long as the supply lines are functioning, and as long as people are able to meet their financial obligations, they do not need to bother much about the entire constellation. They are immediately confronted with it, however, the moment something goes wrong with any of the conditions.

In this way the exploitation of the new sources of energy clearly continues a trend that has always been characteristic of the control of fire. Dependence on the forces of nature has become less direct (which is not to say less in extent!), and at the same time dependence on cultural and social resources has increased. A complicated technical and organizational apparatus is needed in order to have the supply of energy available at all times. Most of this apparatus is located 'behind the scenes' of industrial society, invisible to the ordinary consumer.

The increased supply of fuel has involved an enormous lengthening of the chains of interdependence between groups. An immediate effect has been a great rise in the level of physical comfort. In a modern industrialized country there is normally no shortage of fire. All that people need in order to be able to enjoy the benefits of the abundantly available energy is access to that most general resource of affluence: money. Light and warmth are for sale in unlimited quantities, and so is power to operate a wide variety of appliances. Energy is made available in such a convenient fashion that it is easy to forget the social effort required to produce it.

That social effort is expended, first of all, at the drilling wells and in the mines where the energy is won and, next, during the operations of processing it into consumable gas, oil or electricity, and of transporting and distributing it. The many provisions needed for the undisturbed flow of energy are often taken for granted, but they cannot fail to exert permanent pressure on those who benefit from it as customers. The bills have to be paid, financially and otherwise.

As noted by the German sociologist Peter Gleichmann, the continuous availability of electricity, at every hour, in all parts of the world, has led to a diminution of the contrast between day and night.[21] By the middle of the nineteenth century, the large investments made in their factories impelled many owners to let the engines run day and night. Gaslight illuminated the workplace. In the twentieth century night-life has steadily extended, especially in the cities. Water mains, sewage, gas, electricity, telephone, telefax, radio, police, fire brigade, hospitals – all such services are generally expected to operate day and night. International interdependencies never come to a halt. This is one of the reasons why many people turn on the news as soon as they wake up in the morning: before resuming their daily activities they wish to learn what has happened while they were asleep – in their own country, where it was night, and elsewhere, where it was day.

Once in a while there is a hitch. Sometimes the local supply of electricity is disturbed, as happened in the 'blackout' in New York on 13 July 1977. Or international complications occur, such as the oil crisis of 1973, when the majority of oil-producing countries jointly managed to enforce a drastic increase in the price of crude oil.[22]

But disturbances are remarkably rare, in view of the prodigious growth of energy consumption in the second half of the twentieth century. The industrial economy is a fuel economy, revolving around the regular supply of fuel that can be easily converted into energy. The increase in productivity has led to intensive growth, concentrated in the centres of industrial production and consumption, and to extensive growth, which at present is mostly confined to the rest of the world, the 'periphery'. Even more than extensive growth, intensive growth has, today, all the characteristics of a largely autonomous, self-propelling force. Light, warmth, motion and even coolness are being produced with fuel, in increasingly large quantities. The rising supply of all these fuel-intensive goods and services in turn constantly stimulates demand from customers, who are eager thus to enhance both their material comfort and their social status.

The drive towards intensive growth is also present in those countries that are not yet so highly industrialized. There, however, extensive growth still prevails, and it is only to be expected that in the decades to come this process of extensive growth will increasingly spread to the rich countries as well. Wealth attracts poverty; history abounds with examples of this general rule. Whenever the opportunity offers itself, many people from poorer regions will migrate to regions where people on the average are better off.

Meanwhile, the combined pressures of intensive and extensive growth continue to push up fuel consumption. In most cases, the actual combustion processes are hidden from the consumers, who therefore experience no direct physical inconvenience to deter them from using up increasingly more fuel.

LARGE CITY FIRES

One of the most spectacular ways in which the effects of combustion processes become visible is when a fire 'breaks out' and turns into a conflagration. With the rapid growth of cities in the early modern and the modern era, many huge conflagrations occurred. Thus, as noted before, in London in 1666 more than 13,000 houses were devastated – a record for Western Europe which, however, more than two centuries later, was still being exceeded by fires in the United States. In Chicago in 1871 17,500 houses were gutted; in San Francisco in 1906, following an earthquake, 28,000.[23]

Yet, as I have already noted in Chapter 7, there is no constant correlation between the size of cities and the frequency of conflagrations. As E. L. Jones makes clear, the fire in London in 1666, while exceptionally big, fitted into a general pattern. Like Chicago 200 years later, London was at the time going through a period of rapid and turbulent growth which made it all the more vulnerable to fire. Packed with buildings constructed from cheap and highly flammable materials, it was like a tinderbox. The fire, which had started in the middle of the night in a baker's shop, quickly spread, fostered by the dry summer weather and a strong wind, and it was soon completely beyond human control.[24]

In the nineteenth century the largest city fires in Western Europe raged in Hamburg in 1842 and in Newcastle in 1854, destroying respectively 4,000 and 800 buildings. Cities continued to grow in number and size, but urban conflagrations did not follow suit. In the United States the San Francisco fire of 1906 was the last in a series of increasingly large conflagrations, and although North America continued to have a higher annual rate of fires than Western Europe or Japan, the frequency of big city fires in that part of the world also did not keep pace with urban growth.[25]

In order to explain the 'fire gap', as he calls the discrepancy between urban growth and reduction of conflagrations, Jones points to two factors. First, he notes the increasing use of brick, concrete and steel as building materials. Such rapidly growing industrial cities as Birmingham and Manchester were infamous for their poor housing conditions, yet for the whole of the nineteenth century they were spared large fires. According to Jones, this was mainly as a consequence of the fact that in England, which by this time was largely devoid of forests, it was cheaper to use industrially produced brick as a building material than wood.[26]

The second explanation advanced by Jones concerns spatial arrangements. Especially in North America and Australia, where land was abundant, cities tended to expand in the form of detached single-family houses with gardens. This greatly reduced the risk of fires spreading over an entire area, even if the exterior walls were made of timber and the roofs were covered with rustic-looking thatch.

Jones dismisses a third possible explanation to the effect that the reduction of urban conflagrations might have been the result of more effective fire-fighting. While admitting that in the course of the nineteenth century a series of technical innovations were indeed introduced, he does not think that these could conceivably have enabled the firemen to stop any really serious outbreak of fire.

While I fully accept the first two points made by Jones, I think that he too readily rejects the possible impact of improvements in fire-fighting. It is indubitably true that once a blaze had assumed catastrophic proportions, fire brigades were powerless to stop it. But the problem is precisely this: how to explain the decrease in the frequency of such uncontrollably large conflagrations. More timely and effective interventions on the part of the fire brigades may have played their part.

A closer look at one particular event that was in several respects typical of urban conflagrations in the mid-nineteenth century may help to clarify the issue. On 2 July 1858 a fire broke out in a warehouse on the fringes of a densely built-up district of Amsterdam. Kindled by a strong wind, the flames spread to adjacent buildings – industrial properties and houses – twenty-seven of which were razed completely.[27]

In a historical account of this fire, with quotations from contemporary newspapers, two points are of particular interest. The first concerns the fire brigade, which was made up of conscripted citizens who had been assigned by lot to serve as hosemen. In the case of a fire alarm they had to report immediately, under penalty of a fine. In addition, the commander could also order members of the public to take part in extinguishing operations. The major appliances available were engines of the type designed in the seventeenth century by Jan van der Heyden. In order to induce the hosemen to exert themselves, they were profusely supplied with beer. All in all, as several eyewitnesses noted in letters to the editor, both the technical equipment and the morale of the firemen left much to be desired.

And so, according to these commentators, did the attitude of the public. People flocked in from all over the city to watch the spectacle. In a critical article, written soon after the event, a physician complained about 'the abominable disorder on the premises of the fire, the inferior condition of the material, and the retarded arrival of the civic guard'. What he found particularly exasperating was the fact that the firemen could not make themselves understood,

for their voice was not heard because of the general shouting, yelling and raving in the street, where everyone who felt like it was yelling, raging and shrieking amidst orders and commands which, being drowned in the clamour, were not heard, let alone executed properly. It was more a bacchantic street show, a rowdy popular feast, than a tragic disaster, calling for calm and quiet action.

Both the scene evoked and the indignant responses to it reveal something of the changes that were taking place in fire-fighting. One was technological, as in the decades after 1850 hand pumps were gradually replaced by steam-powered pumps – an innovation that was soon followed by connection of the hoses to the newly installed high-pressure water conduits in cities. No less important for the fire brigades were reforms in organization and discipline.

At first sight, these reforms seemed to go in different directions, for in some cities, volunteer fire brigades were founded, in others, professional brigades. Common to both types, however, was a higher degree of specialization and organization than had existed among their predecessors. The system of recruiting conscripted hosemen was abandoned. Not only did the new brigades have more advanced technical equipment but their members received a better training and were more highly rewarded – either in money and social status, or, significantly, just in status.

The changes were reflected in the tone of newspaper reports of the operations of the fire brigades, highlighting their efficiency and their self-effacing heroism. Fire-fighting came to be held in higher regard than ever before, either as a splendid vocation for citizens who were prepared to put all other chores aside when called for an emergency or as a noble profession for those who devoted themselves fully to this task and were even prepared to offer their lives in its service.[28]

The changing attitude towards the job of fire-fighting, both in the press and within the corps itself, was an aspect of a broader 'civilizing campaign'. The new fire brigades were informed by the same spirit as the societies for rescuing people from drowning, which, during the same period, were founded in coastal towns where for centuries the attitude to the victims of a shipwreck had been callous rather than philanthropic. Until the early nineteenth century the primary, and virtually the only, task of those engaged in fire-fighting had always been to protect physical property; now saving human lives was beginning to gain priority, as evinced by the founding, in London in 1836, of the Royal Society for the Protection of Life from Fire.[29]

Authors writing about nineteenth-century 'civilizing offensives' (or 'campaigns', as they are perhaps better called) usually stress only one side: the attempts, on the part of the bourgeoisie, to educate the working classes in certain middle-class manners. The training of the fire brigades, however, involved far more than the mere imposition of a bourgeois regime on recalcitrant workers; certainly for the volunteer corps, but for most of the professionals as well, it implied self-chosen dedication to a lofty ideal. The civilizing spurt undertaken by the firemen seems also to have affected the public at large, and while big fires continued to attract big crowds, the spectators seemed to be less inclined to treat the scene of the blaze as a fairground.

These considerations, I think, lend support to the idea that, together with the two explanations Jones gives for the 'fire gap', the changing role of the fire brigades also deserves to be taken into account, not just their technical equipment but also their organization and mentality. Nor are the attitudes of the public to be ignored, for as the problem of disorder was reduced, firemen were better able to concentrate their efforts on suppressing the fire.

Today, large urban conflagrations such as used to rage recurrently in agrarian societies continue to take place with some regularity in the Third World. They are indeed so frequent that they receive very little attention in the news media. Even when, in a *barrio* in São Paulo or Manila, thousands of dwellings are gutted, the newspapers in Europe devote no more than a brief note to it, if they mention the incident at all. The very tendency towards 'under-reporting' helps to explain why we are so poorly informed about fires in the past. They were recorded only if they destroyed important public buildings;

normally, the burning down of dwellings did not warrant a mention – especially if, as was usually the case, the dwellings happened to belong to poor people.

In highly industrialized countries, not only has the frequency of large city fires during peacetime diminished drastically but so also has the frequency of fires in general. This is so in spite of the spread of arson for profit, a new cause of fires committed by or at the orders of the owner of a building, in order to obtain insurance payments. While all industrial societies share in the declining trend, some share in it more than others. Each year in the United States and Canada, property losses as well as personal casualties per 100,000 inhabitants are at least twice as high as they are in Western Europe and Japan.[30] Some contrasts are particularly impressive. In the late 1980s Chicago, a city half the size of Hong Kong, had three times as many fire deaths. In Baltimore, about equal in size to Amsterdam, the number of fire deaths was thirteen times as high.[31]

Like fire-related issues generally, these striking differences appear to have failed to attract the attention of comparative sociologists. To explain them, we have to confine ourselves to some *prima facie* reasoning. By far the greatest number of fatal fire accidents in the United States occur in single-family dwellings. It is very likely, therefore, that the high casualty rate is related to the fact that many of these houses are still built of timber. American fire experts also point to other factors such as a lack of formal legislation and of educational programmes aimed at fire-prevention.[32]

While the issue has not yet been looked at from the angle of sociological theory, it is certainly being studied and discussed by a growing number of practically orientated specialists. Many organizations, including fire brigades and insurance companies, take a strong interest in promoting measures to reduce fire hazards. They can rely on a rapidly growing body of scientific and technological knowledge which is being advanced in highly specialized laboratories and promulgated through specialist journals and conferences. The aim of these activities is clear: just as in the past two centuries huge urban conflagrations have become increasingly rare in highly industrialized countries, so the number of smaller fires, which are now the main cause of casualities and material damage, should also be further reduced.

The result is likely to be a strengthening of the general trend towards cultural convergence in the industrial world. There will be continued tension between, on the one hand, the wish for economy and convenience, and on the other hand, the call for safety. In the search for a compromise between these conflicting demands, increasingly uniform rules are likely to be introduced everywhere in the world regarding both the construction and the interior layout and furnishing of buildings. The impact of this tendency can already be seen in airports and in large international hotels. Most fire experts find the existing regulations still far from sufficient; they would, for example, like to see stricter standards relating to the textiles used in furniture and clothing.[33] The ultimate goal is to turn the space in which people live and work into a zone that is optimally protected against fire.

BEYOND THE FIRE-PROTECTED ZONES: WAR

Everything said in the previous section about the decline in urban conflagrations stands in need of one crucial qualification: the decline obtains only in times of peace. With industrialization, productivity has increased, and so, inevitably, has the social potential to destroy, or 'destructivity'. Not only have the means of preventing and fighting fires become more effective, but so have the means of starting fires, while cities as targets for arson have also increased in number and size. Consequently, in the twentieth century acts of war have caused some of the largest urban fires in history.

As noted by the urban historian and sociologist Lewis Mumford, behind their walls, cities originally offered the most secure protection against military violence, but in the age of aircraft and rockets they have become the most vulnerable targets for destruction.[34] Once this had first been clearly demonstrated in the Spanish Civil War (1936–9), air bombardments on cities became standard practice during the Second World War. Their destructive effects steadily increased.

According to the official figures, the German air raid on Rotterdam in May 1940 resulted in a death toll of 980, and made 75,000 people homeless. These figures were exceeded many times during the nightly raids on German cities carried out from 1942 by Allied bombers. The first target, the ancient port of Lübeck, was selected for destruction because it had a late-medieval town centre consisting mainly of highly flammable wooden houses. The desired effect was indeed attained: after a few hours of intensive bombing, 80 per cent of the old city burned down.

In a matter of months the Allied forces established an immense organization to carry out a programme of 'solid destruction, city by city'. For a single night attack on Cologne in May 1942, a fleet of more than 1,000 aircraft was dispatched. The crews were instructed to 'set up a conflagration that no fire brigade in the world has the remotest chance of extinguishing'.[35] Most devastating of all were the bombings, in July 1943 and February 1945, of Hamburg and Dresden. Huge quantities of incendiaries and explosives were dropped on those cities, creating 'fire storms' against which the fire brigades were indeed powerless and from which escape was impossible, so that tens of thousands of people perished through burning or suffocation.[36] Attacks on an even larger scale were launched against Tokyo in a series of 'conventional' bombings in the early months of 1945, and on Hiroshima and Nagasaki by the atomic bombs dropped in August 1945.

Before starting the international war, the National Socialists had already used fire in the civil war they waged to establish the hegemony of their party in Germany. The part they played in 1933 in the burning of the Reichstag is still disputed, but they certainly exploited the fire to their own advantage in a much publicized show trial. Later, they staged *autos de fe* in which books written by Jews or others of whom the party disapproved were publicly committed to the flames; and the infamous *Kristallnacht*, when all over Germany synagogues were set on fire. As early as 1920, troops of German militias (the *Freikorps*) had been buccaneering in eastern Europe, where, with modern means, they practised the old techniques of scorched earth and sacking and burning. The German novelist Ernst von Salomon, in his semi-autobiographical novel *The Despised*, gave a graphic sketch of their mentality.

We made our last push. We roused ourselves once more and charged. To the last man we left our cover and dashed into the wood. We ran across the snowfields and broke into the forest. We took them unawares and raged and shot and killed. We hunted the Letts across the fields like hares, set fire to every house, smashed every bridge to smithereens and broke every telegraph pole. We dropped the corpses into the wells and threw bombs after them. We killed anything that fell into our hands; we set fire to everything that would burn. We saw red; we lost every feeling of humanity. Where we had ravaged, the earth groaned under the destruction. Where we had charged, dust, ashes and charred balks lay in place of houses like festering wounds in the open country. A great banner of smoke marked our passage. We had kindled a fire and in it was burning all that was left of our hopes and longings and ideals. We retired swaggering and intoxicated with success and laden with booty.[37]

Written in a more laconic fashion but equally interesting as a document about the use of fire in modern warfare is the following passage from a straightforward autobiography.

Sir Bindon sent orders that we were to stay in the Mamud valley and lay it waste with fire and sword in vengeance. This accordingly we did, with great precautions. We proceeded systematically, village by village, and we destroyed the houses, filled up the wells, blew down the towers, cut down the great shady trees, burnt the crops and broke the reservoirs in punitive devastation.[38]

These words – which, as far as the tenor goes, might have been taken from the memoirs of any general from Julius Caesar to Napoleon – were written by Winston Churchill. They relate to a campaign of the British army in Afghanistan around 1900. In a matter-of-fact fashion, they bring to our notice the time-honoured policy of arson and scorched earth as a means of intimidation and a display of power. The same message was conveyed, more caustically, in a novel by the nineteenth-century Dutch writer Multatuli, when he mentioned in passing that an Indonesian village 'had just been taken by the Dutch army and was therefore in flames'.[39]

Acts of incendiarism by armies are indeed terrifying and

intimidating. Yet we should not fail to notice that such displays of power also imply a recognition of their very limitations. Fire is used in order to force people to obey when apparently they cannot be made to do so by any other means. A government that has to fall back on such measures in its domestic politics is surely facing a serious crisis of authority. The skirmishes described by Churchill took place at the frontiers of the British Empire, where the *pax Britannica* was not firmly established. Only there did the colonial power regularly resort to incendiarism.

In the early 1980s, also in Afghanistan, Soviet soldiers took to incendiarism. They employed a weapon similar to that which, twenty years before, American troops had been using in Vietnam: napalm. In both cases, the consequences were disastrous for the native population, but eventually the Americans and the Soviets found themselves forced to halt the hostilities and to retreat. Apparently the large-scale destruction by fire was a display of military power that concealed a lack of political power. The same observation could be made again in February 1991, when Iraqi soldiers set fire to the oil wells of Kuwait. Regardless of the strategic justifications given, this was clearly an act of despair by losers; they had to abandon the treasures they had conquered and they consigned them to the flames.

BEYOND THE FIRE-PROTECTED ZONES: FOREST FIRES

People who have grown up in a city, surrounded by fields and pastures under permanent cultivation, can be quite startled when they visit a region of gatherers and hunters or swidden cultivators who regularly set fire to large tracts of land. One of the first people known to have registered his sense of marvel at such a sight was the Carthaginian traveller Hanno, who *c.* 500 BC was sailing along the African coast near Sierra Leone, and saw flames rising high into the sky out of the bush on islands near the mainland.[40] Many other townspeople have since had a similar experience. Thus on the day of his arrival at Port Moresby, New Guinea, in the summer of 1915, the British anthropologist Bronislaw Malinowski noted in his diary:

Fires had been kindled in various places. Red, sometimes purple flames crawled up the hillside in narrow ribbons; through the dark blue or sapphire smoke the hillside changes colour like a black opal under the glint of its polished surface. From the hillside in front of us the fire went on down into the valley, eating at the tall, strong grasses. Roaring like a hurricane of light and heat, it came straight towards us, the wind behind it whipping half-burned bits into the air. Birds and crickets flew past in clouds. I walked right into the flames. Marvellous – some completely mad catastrophe rushing straight at me with furious speed.[41]

As Malinowski was well aware, the awesome fire was not just a force of nature run wild but was started and controlled by the people he was going to study. For them, slash and burn was virtually the only way to clear land for cultivation. Townspeople, on the other hand, not being used to large controlled fires and mindful of how precious wood was, were usually inclined to condemn these practices. They found slash and burn primitive and wasteful, and regarded it as a ruthless form of exploitation. Sedentary farmers tended to think likewise. As far as they knew, slash and burn was practised only on poor soil in remote regions. Most of them would not have realized that at one time their own fields had also been cleared in this way.

In the course of the twentieth century slash and burn has tended to be viewed more favourably.[42] This change of opinion is part of a general change in the cultural climate of highly industrialized countries. Increasingly, people recognize that the most advanced forms of intensive agriculture depend on the continuous injection of enormous amounts of fuel. An awareness of the costs involved in this form of agricultural production has engendered a more positive attitude towards earlier forms.

Along with the changed attitude came a closer interest in the ways slash and burn was actually practised in various circumstances. Thus in 1957 the American anthropologist Harold C. Conklin published a report for the Food and Agricultural Organization on the Yagaw Hanunóo, a community of about 128 people living on one of the Philippine islands. The main crop of the Yagaw Hanunóo was rice, which they cultivated in small plots cleared in the forest. They took great care in selecting a suitable tract of forest. When, finally, they had decided on a particular site, they would first cut the branches from the trees to let them dry in the sun, and then, just before the beginning of the wet season, they would set them on fire. The timing in particular required a delicate touch: if they started burning too soon, there was a danger that the ash would be blown away and valuable substances be leached out of the soil; if, on the other hand, they waited too long, they ran the risk of being overtaken by the rains, which would make burning very difficult. The burning itself was also carried out with great care. The swidden firers worked from a precisely designated perimeter towards the centre; two or three men with bamboo torches kindled the dried-up vegetation at various spots, and then a whole group of men, women and children guarded the fire closely so that it would nowhere move in the wrong direction. Soon after the burning was done, rice and other new crops were sown and planted in the ash-covered soil. After two or three harvests, the field was abandoned and allowed to lie fallow until, after about ten years, the cycle was resumed.[43]

Another positive account of slash and burn practices was published in 1961 by the anthropologist Robert Carneiro, on the basis of fieldwork in a community of about 1,000 Kuikuru Indians in Brazil. According to his findings, a strategy of shifting cultivation judiciously followed would require only a few hours of labour per day to provide people, under favourable circumstances, with 'abundant and reliable' subsistence. For at least three generations, the Kuikuru had been practising slash and burn in a forest area within a radius of no more than 4 miles from their village. There was no shortage of land, so that, when after a few harvests their fields were overgrown by weeds, they could move to new plots and let the abandoned fields lie fallow for up to twenty-five years.[44]

Clearly, the economies described by Conklin and Carneiro were predicated upon an abundance of land. A group could abstain from using an area for ten to twenty-five years only on the condition that it had enough alternative land available. If the population increased, the forest would come under greater pressure – as is nowadays happening all over the world. It is no coincidence that the reports on successful slash and burn economies I have cited date from around 1960. At that time it was perhaps still possible for some small communities of shifting cultivators, in areas far removed from the centres of the industrial world, to practise slash and burn without being exposed to pressures to intensify cultivation and to shorten the periods of fallow.

Such pressures are, of course, almost as ancient as sedentary agriculture and animal husbandry. For a long time, however, in many parts of the world, most notably in Africa, South East Asia and Latin America, shifting cultivators have been able to resist these pressures or to withdraw from them. But in the second half of the twentieth century, intensive and extensive growth have accelerated at such a rate that the conditions for a balanced system of swidden cultivation are becoming exceedingly rare.

Especially in the tropics, huge areas of land are still covered with forest, but each year great inroads are being made. The human population is steadily growing, and through networks of commerce and industry spanning the entire world the same economic forces and the same technical equipment are endangering the remaining tropical forests. If the trees are wanted for timber, they are felled. Just as often, however, it is not the wood but the land that is coveted, and in those cases the colonists usually resort to the most ancient means of destruction: fire.

In *The Primary Source*, the British expert on forests Norman Myers gives a survey of the damage that is annually being inflicted. One of the causes is the need for fuel in the Third World; in many densely populated areas, fuel for the fire to cook on has become proverbially as scarce as the food to be cooked. In addition, there is a rising demand for timber and pulp from the most highly industrialized countries. The greatest threat to the continued existence of the rain forests does not lie in felling the trees, however, but in the indiscriminate burning down of entire tracts in order to clear the ground for raising crops and cattle for commerce.

Around 1980, according to Myers's conservative estimate, about 20,000 square kilometres of forest (mostly in Latin America) were sacrificed to cattle-raising each year, and over 80,000 square kilometres worldwide to agriculture, while another 80,000 square kilometres were seriously damaged. Most of the fires which were destroying the forests were started by small drifting farmers – whose numbers were estimated at 800 million in 1980 – who saw themselves forced to leave their homesteads and to move into the forest. Myers calls them 'shifted cultivators'; he finds that today's typical forest farmer is to be regarded as

an unwitting instrument, rather than a deliberate agent, of forest destruction. He is no more to be blamed for what happens to the forest than a soldier is to be blamed for starting a war. The root causes of his life style lie in a set of circumstances often many horizons away from the forest zones. Far from being an enthusiastic pioneer of forest settlement, he finds himself pushed into the forest by forces beyond his control.[45]

The circumstances indicated by Myers are primarily economic and demographic. The world economy generates a rising demand for products of tropical agriculture. In order to meet this demand, garden plots are converted into plantations, and the small farmers have to leave. At the same time, the population continues to grow, so that pressure on land becomes even greater. As a result, the contradictory situation arises that in a world suffering from severe shortage of wood, each year many tens of thousands of hectares of forest are set on fire.

Myers's book contains some satellite photographs taken at night which show the rain forests as light spots, illuminated as brightly as large cities. An eloquent testimony in words to the ubiquitous conflagrations is the following passage from the Ghanaian anthropologist Jon Kirby.

It is hard for one who has not seen it to visualize the utter destruction caused by a bushfire in the heat of the dry season. Crackling dry vegetation is turned to cinder in seconds and hardwoods such as the acacia and shea-nut smoulder for days before toppling. The aftermath looks like the Maginot Line or an eerie moonscape. And the effects seem to be getting worse each year.[46]

According to Kirby, the demographic and economic causes of setting fire to bushes and forests are reinforced by cultural motives. Sedentary farmers traditionally regarded the forest as an alien and hostile force, the abode of vermin and evil spirits. The memory still lingers of the times when their own fields were carved out of the bush, and they never forget to rule out the possibility of the bush one day growing back over their gardens again. Therefore, wherever they have the chance, they try to suppress the bush by burning.

This deeply rooted mentality makes the farmers impervious, in Kirby's view, to rational ecological arguments. Burning the bush is for them not merely an economic operation; it is hallowed by a deeply ingrained sense of necessity and mission.

The household has had to reclaim land from the 'wild' and drive its forces back. It has employed every available means to make war on it, especially fire, which thins it out, makes it acceptable, squeezes it for meat, honey and chases the wild animals away. From time immemorial, fire has been a wall of culture against the 'wild' which has been thought of as inexhaustible, evil, dangerous, unknown and useless for any socialized purposes. It has been man's primary ally in the constant work of domesticating the 'wild'.[47]

Kirby's words suggest an attitude the origins of which may have to be sought in a very distant past, long before the emergence of agriculture. His interpretation is in line with the mythological evidence collected by Sir James Frazer and Claude Lévi-Strauss, to which I referred in Chapter 1. A more down-to-earth suggestion in the same direction is made by the Australian historian Geoffrey Blainey, who notes that to set fire to a dead bush or a tussock of dry grass 'often confers a stronger sense of omnipotence than its modern counterpart – driving a powerful car or motor-bike'.[48] Yet before attributing to humankind a primordial, deeply rooted urge to burn the bush, we have to realize that any such urge is always shaped by cultural traditions. The descriptions of swidden agriculturalists such as the Yagaw Hanunóo and the Kuikuru suggest that people can also develop a more pragmatic attitude towards fire, using it on a large scale, but not at random. The almost obsessive desire to burn away the bush is perhaps more indicative of a particular stage in the development of agrarian society, when recently settled farmers found themselves compelled to defend their fields against the wilderness still surrounding them.

There can be no doubt that at present many peoples relish burning the bush. As noted by the development worker Albin Korem, children in Ghana are encouraged at an early age to adopt the habit of setting fire to the bush, and to enjoy it.[49] There seem to be hardly any countermanding forces to restrain them and to curb their delight in bush-burning.

The Dutch anthropologist J. M. Schoffeleers studied the demise of such countermanding forces in Malawi in Central Africa. There, according to his account, fire-management was an art which had locally been 'developed to perfection'. Since fire, 'in the hands of the untrained and irresponsible', might cause great damage to a community's resources, its use was subject to 'stringent legislation and severe sanctioning'. The control was exerted by the priests of the regional cults. They imposed rules both with regard to the season in which burning was allowed and to the areas that could and could not be burned.

The season was opened with the ceremonial burning of a hill by the priesthood of the Bunda shrine in the first week of September and no burning was allowed before that date ... Large tracts of forest and brushland were protected from burning by means of ritual interdictions. Transgressions in either case were thought to be a major cause of drought and punishments were severe. It is probably no coincidence that some of Malawi's largest forest areas were destroyed by fire in the second half of the nineteenth century, when a combination of factors ... led to an almost complete collapse of the territorial cults.[50]

In Malawi and Ghana, as in many other African countries, burning practices are no longer restricted by priestly sanctions, which at one time were severe indeed, and included the penalty of selling an offender into slavery. At present, according to Kirby, the state authorities in Ghana tend to underestimate the severity of the problem of virtually uncontrolled burning, seeing the shortage of water as their country's greatest ecological problem. However, Kirby argues, the water shortage is to a large extent caused by the wanton use of fire. Burning the vegetation results in deforestation, deforestation in erosion, and erosion in desiccation of the soil and declining rainfall.[51]

The same chain of events may be observed in many other countries.

The process of deforestation, which first affected the temperate zones, is now taking place in the tropics, at a much faster pace and on a far larger scale. The current development is especially alarming since the tropical rain forests form closed ecosystems, in which practically all the available nutrients are absorbed in the living biomass. When a section of rain forest is burned down, some of the nutrients immediately vanish in vapour and smoke, and the remainder are reduced to ashes. If, subsequently, the ashes are washed or blown away by rain or wind, those too are lost to the ecosystem. The forests have been able to survive small-scale burning as practised by the Yagaw Hanunóo or the Kuikuru Indians, but they cannot withstand the present rate of destruction by fire.

Besides the fires set for clearing land in the Third World, countless bush and forest fires are each year ravaging the remaining 'nature areas' in the rich parts of the world. Every summer the media report conflagrations in southern Europe and in the United States in which many tens of thousands of hectares are destroyed.[52] Sometimes the cause of ignition is lightning, but just as often it is human negligence or intent.

An increasing number of ecologists believe that, whatever the immediate cause, in most cases the actual size of the conflagrations is an effect of previous human intervention. Among the largest bush fires of recent years were the 'Ash Wednesday fires' in the State of Victoria, Australia, in 1983, in which seventy-two people died. An Aboriginal elder was later quoted as saying that the catastrophe was permitted to occur because the country had not been 'kept clean'.[53] This statement expresses a view, one gaining wide support among foresters, that an unduly anxious concern to protect the forests against fire is likely to have an adverse effect.

Forceful statements expressing this opinion are to be found in the volume *Fire and Ecosystems*, published in the United States in 1974. The authors unanimously condemn indiscriminate, 'promiscuous' burning, but they are equally critical of 'fire-exclusion programmes' which, in their view, are informed by prejudice and ignorance and by unwarranted fear of fire in any form. For more than 100 years, conservationists have considered fire as 'an insidious enemy' and, acting on that idea, have initiated measures aimed at banning fire altogether from their forests. The result has been an accumulation of dead trees and litter which, once ignited, can easily spark off immense conflagrations: 'The fuels continue to build up and become more widespread, and when fires do get out of control, the toll in human life, natural resources, and costs is enormous.'[54]

The authors therefore urge that people 'relearn the lost art' of using fire as 'a servant' and 'a useful friend'. They remind us that 'fire is a bad master, but, when employed correctly, it is a good servant', and they advocate the 'skilful application of fire as a management tool'. Enlightened fire policies are needed, grounded on the principle that 'underburning' will eventually result in 'overburning'. From time to time, 'prescribed fires' will be necessary to clear away 'fuel build-ups'. Thus the book commends 'an allowance for fire to follow its natural course under carefully specified conditions'.[55]

There is a striking similarity between this ecological strategy and a more general tendency towards conscious permissiveness which may be observed in highly industrialized societies. This tendency has found its most explicitly articulated expression in certain theories of mental health, inspired by psychoanalysis, which are based on the principle that any attempt at wholesale suppression of natural human urges will be futile and counter-productive. It is a tendency which Norbert Elias, in the context of his theory of civilizing processes, has identified as a trend towards a 'controlled decontrolling of emotional controls'.[56] In a similar vein the indiscriminate suppression of forest fires is now deliberately tempered, as a part of the process of continuously extending human control over fire.

9. THE CONTROL OF FIRE AT DIFFERENT LEVELS

══

THE INDIVIDUAL ACQUISITION OF CONTROL OVER FIRE

There are estimated to be over 5 billion people living on our planet today. They use fire in a multitude of different ways, depending on the society to which they belong and on their position within that society. Some spend many hours a day collecting firewood and carrying it home; others have enormous quantities of energy at their beck and call. What they all have in common is that, directly or indirectly, they use fire and they need fuel.

As mentioned before, certain plants and trees are fully adapted to frequent exposure to fire, requiring periodic fires for their survival and reproduction – for example, because their seedpods will open only at high temperatures. In the process of natural selection they have become 'pyrophytes', or 'fire-growers'; without fire they would be either overgrown by competitors or unable to reproduce themselves.

For many thousands of generations, humans have been living in conditions in which they have needed fire for their continued existence and propagation. While this may justify calling them pyrophytes too, in their case the label is no more than a metaphor. There is no feature in their biogenetic equipment that makes humans dependent on fire, the way they are, like all land animals, dependent on the other three 'elements' of classical cosmology: earth, water and air.

Still, the experiences of tens of thousands of generations in dealing with fire may have left some traces in the genetic structure of humankind today. I am told that horses and other steppe or savannah equines have evolved a very specific reaction to grass fire: rather than fleeing, they advance towards it and jump over.[1] While this might initially strike one as surprising, it is probably a far more effective strategy for survival than running away in front of a fire that is rapidly sweeping over the grassland. It is also likely that, during the lengthy Palaeolithic phase, a process of 'fire-selection' occurred among our hominid and human ancestors. Those individuals who were better adapted to a socio-cultural fire regime probably had greater chances of survival and reproduction than those who were less well equipped in this respect.

This is not to say, however, that the increased control of fire – especially during the last few centuries – is to be attributed to a biological mutation, as a consequence of which people's inborn aptitude for handling fire was suddenly raised. We are dealing here with a process of socio-cultural development. As noted by the American social psychologist Leon Festinger, in such processes the mediocre can profit from the bright; once an invention has been made, other people do not need to resolve all the difficulties that the original inventor had to face.[2]

This principle clearly applies to human control over fire. Ever since the original domestication took place, control over fire depended primarily on social organization and cultural tradition. In each generation people had to learn anew how to adjust to the presence of fire. They had to regulate both their mutual relations and their individual impulses and feelings in such a way as to ensure the regular possession and use of fire. In order to subdue fire, they had to subdue each other and themselves.

The observation that the technical mastery of fire rests on social conditions has been put forward convincingly by Catherine Perlès in her writings about fire in prehistory. As noted in Chapter 2, the same insight has been applied to our own time by Gaston Bachelard. He pointed to the fact that control of fire always involves social authority. A child is first introduced not just to 'fire' but to 'social fire' – fire surrounded by signals given by other people. Even the fear of fire, which may strike us as spontaneous and natural, has been preceded by social experiences: by warnings and prohibitions, by admonitions to be cautious and to keep away from the fire. Hence Bachelard's conclusion that fire is to people 'more a social reality than a natural reality'.

The social prohibitions are the first. The natural experience comes only in the second place to furnish a material proof which is *unexpected* and hence too obscure to establish an item of objective knowledge. The burn, that is to say the natural inhibition, by confirming the social interdictions, thereby only gives all the more value to the paternal intelligence in the child's eyes.[3]

These comments are highly perceptive, but, as observed before, they leave out one aspect: socio-cultural development. The prohibitions alluded to by Bachelard are the prohibitions issued by parents living in flammable cities in flammable houses filled with flammable belongings. They represent a rather recent stage in the development of the human fire regime.

The fire regime, the complex of socio-cultural commands and options with regard to fire, has changed in the course of time. In his classic study Norbert Elias drew attention to a sentence from a fourteenth-century manner book: 'Thingis somtyme allowed is now repreuid' – things that were once allowed are now forbidden.[4] This, he noted, may be read as an appropriate summing up of a tendency in the civilizing process that may at times be clearly dominant. And it is precisely what happened in the fire regime when, in the transition from gathering and hunting, or from slash and burn, to settled agriculture and urban life, the licence to burn was severely checked.

As I have argued in Chapter 1, our hominid ancestors first encountered fire in the wild before they learned to make it themselves. The same sequence is still repeated in each individual: a child first sees fire burning, and only later does it learn to make fire. The fire that it sees burning, however, is very rarely a 'wild' fire; most children make their first acquaintance with fire in a domesticated, controlled form.

Ever since the original domestication of fire was achieved, all successive generations have grown up in a group-with-fire. Becoming a member of the group implies joining the fire regime – that is, acquiring the knowledge and skills needed to be able to handle fire and to act properly in its presence, in accordance with prevailing group norms.

Clearly, in learning to control fire, individuals do not have to repeat within their own personal history the entire process of socio-cultural development. On the contrary, they have to adapt to the level at which society finds itself during their lifetime. Thus in the last decade of the twentieth century children do not need to learn how to make fire with wooden implements or flints; most of them do not even need to acquire the skill of keeping a wood fire or a coal stove burning. But if a child growing up in a city has any impulse to start large fires, it has to learn to suppress that impulse. For thousands of generations, setting fire to tracts of land was not only allowed: it had positively valued functions. Nowadays, the way of handling fire that would have been normal and useful in a gathering and hunting or a slash and burn economy is considered pathological and criminal.

Until recently, as soon as children started to move around there was a strong possibility in most communities that they would come near a fire. Among gatherers and hunters this would be an open campfire. Thus the anthropologist Jane Goodale described how among the Tiwi of Melville Island in North Australia a couple of small girls aged about two and three were left alone near a small fire while their mothers were away on a yam-digging expedition. After a while, the children decided to build their own fire:

They gathered a small heap of grass, collected a glowing stick from their mothers' fire, and carried it to the heap of grass. They held the glowing stick to the grass and then, lying on their stomachs, blew gently till a flame appeared. Then they scurried about trying to find enough small twigs to feed the fire, but it died out.[5]

Jane Goodale never heard Tiwi parents forbidding their children to play with fire, nor even admonishing them to be careful. Apparently the maxim 'experience is the best teacher' was rigidly followed.[6] A slightly different picture emerges from reports about the !Kung San in southern Africa. There, too, children were often seen picking up embers or burning branches from an open fire, but they were warned to be careful, and, despite these warnings, injuries occurred 'with an unsettling frequency'.[7] The anthropologist Lorna Marshall was present on two occasions when little children, whose mothers had taken their eyes off them for a few minutes, picked up burning sticks from the open fire, dropped them on the soft, dry, bedding grass in a shelter and, at the first burst of flame, sensibly ran outside unscathed.

On the first occasion, the two children, who were about three years old, were frightened and were soothed and comforted by their mothers and other relatives. They were not scolded. On the second occasion ... [a two-year-old girl] had set fire to her grandparents' shelter. She was not apparently frightened at all and was found placidly chewing her grandfather's well-toasted sandal. She was not scolded either.[8]

Small children growing up in a modern urban environment are less likely to come into contact with an open fire. But in many households they may soon find within their reach matches or other means of starting a fire. At that point it is imperative that they learn to handle these very cautiously, to avoid either burning themselves or causing injury or damage to others. The ease with which a match can be lit bears no relation to the destructive potential of the fire thus started. Learning to handle matches is therefore an integral part of the individual civilizing process in a modern society.

As yet, educationalists and psychologists have paid little attention to this aspect of personal development. The subject of fire control is hardly ever mentioned in the standard textbooks on education, developmental psychology, cognitive psychology or social psychology. Apparently it is taken for granted – as it was taken for granted in ancient Israel and Greece and Rome – that children receive sufficient training in the use of fire from their parents or their peers. This, however, is becoming increasingly questionable.

Thus although fire continues to be an integral part of modern society, like society at large it has become highly specialized, and most of its functions are performed 'behind the scenes': in power plants and factories, or in the boilers where the water for radiators and taps is heated. It has sunk deeply into the 'infrastructure' of society. Consequently, for many people, children and adults alike, exposure to actually burning fire may be confined to special occasions at which candles or a log fire or, more rarely, torches are lit for decorative or ceremonial purposes.

Regular everyday use of fire is becoming increasingly rare. One of the few forms in which fire continues to be used directly, and the main reason why many people carry matches or lighters, is smoking. In recent years, many 'civilizing campaigns' have been directed against this habit. Smoking is known to be a major cause of fires; the current campaigns, however, are primarily concerned with health. They appeal to people to exercise self-restraint in order to reduce the risk of cancer. Unintentionally, the campaigns seem to have been successful mainly among the middle and upper classes; perhaps, from there on, an equally unintended 'trickle down effect' will occur, causing this almost vestigial use of fire also to disappear.[9]

The gradual elimination of fire from everyday life has led to a contradictory trend: while the societal capacity to control fire has been increasing spectacularly, the average individual competence in handling fire is probably waning. Clearly, in the highly industrialized rich societies, all sorts of specialists have acquired an unequalled professional expertise in dealing with fire. Some, such as stokers and welders, find themselves confronted with flames and heat in their daily work. Others are engaged in such divergent pursuits as constructing steam turbines for electrical power plants, building rocket engines for launching space shuttles, or conducting experiments with nuclear fusion; although their work results in the sophisticated manipulation of highly concentrated fire, they themselves are in no way directly exposed to it. The professional group that has the most immediate contact with fire is probably the one we continue to call the 'fire brigade'; that profession now also includes highly specialized technicians who are experts in preventing and extinguishing fires in chemical factories and oil fields.

The expertise of specialists is matched by far-reaching ignorance and impotence on the part of non-specialists. People who happen to live in the vicinity of a chemical factory or a nuclear power plant can do very little to arm themselves against the poisonous fumes or the radiation that might be emitted in the event of a fire; should that happen, they could only allow themselves to be evacuated. This may be an extreme case, but even with regard to domestic fire hazards, most people are ill-informed and ill-prepared. Houses, furniture, carpets and curtains, clothes, cars – the average consumer is unable to judge to what extent the synthetic materials these objects contain are flammable. All he or she can do is read the instructions, if any are given, and trust that the manufacturers have observed the safety regulations, again if these exist and are sufficient.[10]

The continuing differentiation of the fire regime is reflected in the extent to which fire and the use of fire are made into the subject of specialized studies. In the general theories of the natural as well as the social sciences, the concept of fire is largely absent. But in both fields, practically oriented studies are burgeoning. While, as noted before, psychology and education textbooks still ignore the problem of how children learn to cope with fire and with the many formal and informal regulations surrounding it, there is a growing body of empirical studies, designed mainly with the practical aim of promoting fire-prevention.

As such, these studies themselves form part of a 'civilizing campaign' that is directed at strengthening individual behavioural controls. Thus the American psychologist Ditsa Kafry investigated 'fire behaviour and knowledge' among six-, eight- and ten-year-old boys in the express belief that cooperative research would 'decrease the hazardous and agonizing use of fire and ... enhance its proper use for warmth and pleasure'.[11]

Almost half of the ninety-nine boys in Berkeley interviewed by Kafry told her that they had played with fire, and one out of five had caused a fire – most of which were easily extinguished and were never reported to the fire department. The accidents caused by the youngest children could be attributed to incompetence in handling the fire; in the case of the older boys, however, proneness to starting fires seemed not to be related primarily to deficient skill or knowledge but rather to a personality trait described alternately as 'rascality' or 'lack of impulse control'.[12] As part of a more general tendency towards mischief, these boys also failed to comply with the fire regime;

in terms of psychoanalysis, their 'ego' was 'not capable of dealing adequately with the reality situation'.[13]

In modern urban society the 'reality situation' (or the fire regime) requires that people do not cause fires. This is clearly in the common interest. Everywhere, by virtue of their possessions, people are hostages to fire. They have therefore every reason to fear and condemn arson in any form.

What is feared in particular is a persistent tendency towards incendiarism, widely known as 'pyromania'. Even though most psychiatrists have come to doubt the usefulness of that term, it continues to be popular. The opposite, 'pyrophobia', has never caught on, although the psychiatric literature does contain sporadic references to 'phobic aversion to fire'.[14] If a person displays symptoms of excessive fear of fire, these are far less likely to arouse concern than does proneness to arson.

While certain individuals do show a habitual tendency towards starting fires, psychiatrists have tried in vain to isolate a clear-cut 'firesetter syndrome'. According to current specialist opinion, incendiarism can be considered, along with other forms of antisocial behaviour, as indicating 'a general lack of self-control, self-confidence, and the skills, particularly the social skills, necessary to obtain rewards from the environment in an appropriate manner'. Starting a fire, it is suggested, 'may offer a kind of control over the environment which the firesetter has been unable to obtain in other ways'.[15]

In the introduction to their monograph on *Pathological Firesetting*, the American psychiatrists Nolan Lewis and Helen Yarnell give an arresting description of the fascination that incendiarism may exercise.

Being an *excellent* means of destruction, it [fire] is *admirably* suited to the carrying out of aggressive tendencies, for the venting of hatred and a *perfect* medium for the discharge of a considerable volume of other suppressed emotions ... By the use of a match the firesetter achieves tremendous spectacular effects which transcend the usual proportion between effort and result. He feels that he has accomplished what the forces of nature released by him are producing for him.[16]

Their choice of adjectives leaves no doubt that the authors have an open eye for the temptations of arson. They also point to the symbolic meaning of fire, as the pre-eminent agency for destroying evil. The ability to recognize this symbolic meaning can be acquired through social learning. The idea that fire 'purifies' is an element of culture, promulgated through various channels, including religion, literature and film. Some of the most esteemed and most popular novels and films of the twentieth century end with a spectacular fire climax – a cathartic scene in which the main character sets fire to his or her house and perishes in the flames.[17]

The individual psychopathology of firesetting may thus be seen to reflect a more general ambivalence towards destruction, and especially towards destruction by fire. This ambivalence seems to be present in every culture. It is probably as old as the domestication of fire itself. The main reason why our early ancestors went to the trouble of incorporating fire into their groups was that they could use its destructive power to their advantage, in clearing land and in cooking. Fire was cherished as a means of destruction that could be turned to purposes of production and protection.

Such concepts as 'production' and 'protection' are tricky because they leave open the issues of just who is producing what for whom, and of just who is protecting whom against whom. The problem of firesetting arises when individuals turn the destructive power of fire to their own – real or imagined – advantage, against the interests of others. In so doing, individual arsonists (or, for that matter, arsonist gangs) follow a practice that at a higher level of social organization may not so readily be considered pathological. Ever since the emergence of agriculture and the establishment of villages and cities, human groups at war have resorted to burning enemy property. In the twentieth century military incendiarism has reached unprecedented proportions. Newspapers and news reports on TV provide the public almost daily with new examples of arson committed in organized conflicts, often in the name of lofty political ideals. Analogous to the – largely unsolved – problem of individual firesetting, there is an immense – and equally unsolved – problem at the level of society at large.

Arson by individuals is generally regarded as a serious issue. The same cannot be said of another type of activity related to fire: the indiscriminate use of fuel. While in the richer parts of the world the direct use of fire is diminishing, fuel consumption continues to rise. Increasingly higher numbers of individuals are getting more energy put at their disposal, with increasingly less effort. They are becoming more and more used to even temperatures and omnipresent light, and to numerous other amenities of a fuel-intensive economy. The growing demand for material comfort and luxury, fostered by the apparent profusion of easily obtainable energy, shows all the marks of limitless insatiablity of the kind characterized by Emile Durkheim as 'anomie'.[18]

VARIATIONS IN FIRE USE AMONG AND WITHIN SOCIETIES

It is a central thesis of this book that the monopoly over the use of fire has contributed greatly to increasing the differences in behaviour and power between humans and other animals – a process that was already under way during the lengthy first phase in socio-cultural development, before the emergence of agriculture and husbandry. During this phase, differences in behaviour and power within societies were mostly determined only by age and sex, while over long periods of time the cultural repertoires of various groups tended to be highly similar.

After the emergence of agriculture, the process of increasing human dominance over other animals continued. In addition to it, however, there also emerged a strong tendency towards increasing differentiation in behaviour and power within humankind itself. Thus socio-cultural development over the past 10,000 years was marked, in the terms of the Brazilian anthropologist Darcy Ribeiro, by an interplay of both 'homogenizing' and 'diversifying' trends.[19]

On the one hand, the gradual expansion of agriculture exerted similar pressures everywhere; this is what the Romanian historian of religion Mircea Eliade alluded to when he observed, 'With the discovery of husbandry ... man was destined to become an agricultural being.'[20] On the other hand, agriculture also gave rise to increasing diversity – among those societies with and those still without agriculture; among societies concentrating on such different staple crops as wheat, rice and maize; and among the various social classes that emerged in agrarian society.

In considering the resulting panorama of human cultures, it may be more tempting to focus on the many variations than on the similarities. The more nearly unique and possibly bizarre the variations, the more likely are they to catch our attention; whereas it is in the very nature of similarities and regularities to be monotonous and to appear self-evident and trivial. The fascination with striking peculiarities in behaviour and culture is already clearly manifest in the writings of Europe's most ancient historian and anthropologist Herodotus, and it has not lost its strength today.

Thus a standard item mentioned in texts relating to the control of fire is the alleged fact that among all peoples known to modern anthropology, there was one that did not know how to make a fire: the inhabitants of the Andaman Islands in the Indian Ocean. This piece of information, which has been cited over and over again, goes back to the British anthropologist A. R. Radcliffe-Brown, who in 1922 published a lengthy and authoritative monograph on the Andamanese. He wrote:

The Andamanese are perhaps the only people in the world who have no method of their own of making fire. At the present time they obtain matches from the Settlement of Port Blair, and a few of them have learnt, either from Burmese or from Nicobarese, a method of making fire by the friction of pieces of split bamboo. Formerly, however, they had no knowledge of any method by which fire could be produced. Fires were and still are carefully kept alive in the village, and are carefully carried when travelling. Every hunting party carries its fire with it. The natives are very skilful in selecting wood that will smoulder for a long time without going out and without breaking into flame.[21]

In assessing these words we have to realize that they were based on observations made many generations after the arrival of the first Europeans, and several generations after the invention of the industrially manufactured safety match. The fact that by 1920 no one on the Andaman Islands knew how to make a fire without matches is not amazing; the same would have been true for Majorca or the Isle of Man. The observation in itself would certainly not be sufficient grounds for concluding that the Andamanese were the only people on earth unable to make fire.

Yet Radcliffe-Brown's words have been taken up eagerly in the secondary literature, without his qualification 'perhaps'. This does indeed appear to point to a desire to relate unique cases, based on the assumption that the extraordinary is more interesting than the ordinary. I would hold, however, that extraordinary cases are really interesting only if we are able to explain them in a way that sheds new light upon the general pattern. This angle is altogether lacking from the story about the Andamanese, which is no more than a curiosity, a trivial 'exception' that is supposed to 'prove the rule'.

Actually in all known societies there were at least some people who knew a method of making fire. In other respects, too, the fire regimes in different societies showed many similarities, the most important variable being the level of agrarian and industrial development. Thus when we look at the use of fire for illumination, we see that over a long period of time only a few changes occurred, allowing for some variation. Even after the introduction of agriculture and the emergence of cities, the domestic hearth continued to be the main source of light, supplemented by torches, oil lamps and candles. Lamps and candle-holders in different cultural areas such as China, India and Western Europe looked very different, but the differences were of style not structure. Although embellishment varied, the technical principle of making light was the same.

But within each of the major culture areas differences in lighting loomed large. In order to illuminate their houses and palaces, the wealthy ruling classes were able to charter specialists who could supply them with lamps and fuel of the best quality. Around 1500 the upper classes in Western Europe adopted the habit of burning wax candles, and this allowed them to stay up till late in the evening – an expensive habit that only a small minority could afford.[22] They shifted the moment of getting up, of meal times and of going to bed to progressively later hours. It was part of the transformation of manners described by Norbert Elias in *The Civilizing Process*. Thanks to their lavish use of candles, the upper classes were able to distinguish themselves not only by what they did, and how, but also by when they did it.

Status competition among the élites probably stimulated new technical discoveries. The growing need for luxury (which was hard to distinguish from the need to display luxury[23]) created a favourable climate for innovations in lighting and heating and in various fire-sustained arts of producing goods that could make life more comfortable and 'richer'. Thus the basis was laid for the mass production of luxury articles in the nineteenth and twentieth centuries.

In stratified societies the unequal distribution of power has always been clearly reflected in the different degrees to which various groups of people were able to use fire and to which they were exposed to the risk of conflagrations. Some groups even ran, or run, an increased risk of being put to death by fire. The latter could be the fate of war casualties, but it could also affect people during 'peacetime', as during the persecutions of heretics and witches in late medieval and early modern Europe. The power of a tightly organized group over individual victims was acted out dramatically in the protracted and extremely painful torture by fire which, according to seventeenth- and eighteenth-century reports, some Indian tribes in North America inflicted upon their prisoners of war.[24] A contemporary example is the practice of widow-burning in India, where women find themselves in a particularly vulnerable position after their husbands' death, given the extremely uneven balance of power between the sexes. Today this practice is officially banned; it would be interesting to compare its social context with that of witch burning in early modern Europe and North America.

The prevailing differences in wealth and power among and within societies clearly affect people's fire use and exposure to fire hazards. In cities all over the world, slum areas and shanty towns are far more likely to be hit by severe conflagrations than the more affluent residential districts. The Bronx in New York is a notorious example. In one of his essays, Salman Rushdie gives a poignant report of 'an unimportant fire' in a slum in London in 1984:

When it started, no alarm rang. It had been switched off. The fire extinguishers were empty. The fire exits were blocked. It was night-time, but the stairs were in darkness, because there were no bulbs in the lighting sockets. And in the single, cramped top-floor room, where the cooker was next to the bed and where they had been housed for nine months, Mrs Abdul Karim, a Bangladeshi woman, and her five-year-old son and three-year-old daughter died of suffocation.[25]

The differences in fire hazard between various districts in New York or London reflect, at a local level, the most important contrasts underlying variations in the use of fire in the world today – the contrasts between those countries in which the majority of the population lives in luxury and those in which the majority lives in poverty. These contrasts are expressed even more clearly in differential access to fuel and to appliances which consume fuel. Thus around 1985 the average citizen of the United States consumed forty times as much energy as the average citizen of India.[26] This figure has to be assessed, of course, in the light of the age distribution of the respective populations, but even so it gives us an indication of the differences in life chances between the highly industrialized rich countries and the less highly industrialized poor countries – among which India is far from being the poorest.

Despite the contrasts which divide it, humanity is becoming increasingly more integrated into one global society. The tensions in this global society are regularly released in violent conflicts: wars, civil wars, revolutions. In Chapter 8 I have given some examples showing the great extent to which fire continues to play a central part in even the most technically advanced wars. It is also still a frequently used weapon in conflicts on a smaller scale.

Sometimes burning appears to serve mainly ceremonial or even theatrical purposes. Thus in August 1989 the democratic government of Greece staged a public event at which the archives of the secret police were burned. It was a festive occasion, in the course of which thousands of files went up in smoke. Similarly, every once in a while there are reports of governments burning large stocks of confiscated marijuana. In all such cases, the fire obviously has a symbolic purifying function: something that is hated or condemned is being destroyed.

But the symbolic burning suits a more practical end as well. What else could the authorities do with falsely incriminating papers or with contraband drugs but destroy them? As in more traditional burning sacrifices and fire festivals, destruction by fire seems to be an emotionally satisfying solution to the problem of what to do with material goods which cause embarrassment to their possessor.

The same intertwining of symbolic and utilitarian functions can also be observed in the use of fire during public demonstrations and riots. A barricade of burning vehicles and tyres can be an effective means of blocking a road, although it is unlikely to stop heavily armoured tanks. But the flames and smoke are also intended to convey a message, and most demonstrators know that pictures of the fire are likely to be shown on TV, as a widely broadcast testimony to their anger and political will.

From burning roadblocks to actual arson is but a small step. During riots, public buildings – law courts, police stations, tax offices, headquarters of ruling political parties – form a favourite target for arsonists. Compared with wars between states, the damage is usually small, but the dramatic impact may be great, since the state's authority has been openly defied and flouted. Here lies the threatening force of such slogans as 'The fire next time' and 'Burn, baby, burn'. As an activist in Brixton was quoted saying: 'They didn't know there was a problem here before we burnt the place down. Maybe we need another one [fire] to show them that things ain't much better now.'[27]

In this sort of conflict, fire is especially popular as a weapon for those who have no access to the state monopoly of organized violence. While they may flinch from murder, they are prepared to commit arson as a penultimate act of violence. Some groups, however, use fire not only against material property but also against people. Thus lynchings in the Southern states of the United States were usually carried out with fire, as are the 'necklace murders' in South Africa – ritual executions in which the victim is killed by hanging a tyre around his neck, pouring petrol over it and setting it alight. Possibly even more dramatic are the instances of individuals who, in order to draw attention to what they experience as unbearable injustice, burn themselves in public.

Pictures of all such events appear on television and in the newspapers. Not a single day passes in which fire is not to be seen on the TV news. And almost always the fire spells violence, unrest, anger, chaos, destruction. Outbreaks of fire, and their consequences in the guise of burnt-out houses and scorched wrecks of cars and buses, form an almost daily expression of collective hatred. Safely nestled down in their chairs, viewers may shudder at the spectacle and hope that the symbolism of fire will continue to be brought home to them only indirectly, via the TV screen.

INCREASED CONTROL OVER FIRE FOR HUMANITY AS A WHOLE

Since the beginnings of the domestication process, the human capacity to make use of fire has increased enormously. At first, for many thousands of generations, the rate of change was low. But then, after the emergence of agriculture, a more rapid succession of innovations occurred, first with the introduction of pottery and metallurgy, and then with the development of more and more specialized techniques. Only about ten generations ago did industrialization begin to become a dominant trend, and with it the capacity to control fire increased at a progressively higher rate.

From early on, by adding fire as a non-human source of energy to their own physical power, humans enlarged their life chances. Increasingly, they came to differ from other, related, animals – in behaviour and power. This led, in the long run, to a prolongation of the average lifespan and a rise in material comfort (or rudimentary 'intensive growth') and, concomitantly, to an increase in human numbers ('extensive growth'). Control over fire was not the sole cause of the process of increasing human dominance, but it played an integral part, and helped to give it momentum.

Industrialization has brought far-reaching advances in the control of fire. A clear example of the overall trend is, again, the development of lighting. For seventeenth-century England, it could still be said without much exaggeration that the fireplace was 'the source of

warmth and the main source of light after darkness had fallen'.[28] This
situation was to change radically in the next few centuries, first by a
series of improvements in the manufacturing of candles and oil lamps,
and then by the introduction of gaslight and electric light.

Gaslight brought unequalled possibilities for illuminating houses
and streets. It was installed in many cities, at great expense, but within
a few generations it was superseded everywhere by electric lighting.
Over the past 100 years the electrification of lighting has been a
dominant trend all over the world. In its initial stages great inventors
and entrepreneurs such as Thomas Edison earned fame by making
major contributions, but from the outset the process had a momentum
of its own. We could equally well say that Edison was propelled by
this very trend, which confronted him with the challenge to stay
ahead in a continuous race of technical and economic competition.

Today, the inhabitants of highly industrialized countries have a
regular flow of electricity at their command so that, at every moment
of the day, they are able with a minimum of effort to produce light.
Compared with the care and skill required for handling a candle or
even a gas lamp, electric light is a most undemanding convenience.
This makes it easy to understand why it has been adopted so rapidly
all over the world. The brief period when there are communities
with and communities without electricity is rapidly coming to an
end; global electrification is an example of the general reduction of
differences among and within societies.

As such, it fits the expression Norbert Elias has used to characterize
an important aspect of the civilizing process in Europe in the twentieth
century: 'diminishing contrasts, increasing varieties'.[29] Electricity has
opened up a broad spectrum of activities; new applications continue
to appear in rapid succession, especially since the rise of micro-
electronics and the ensuing 'automatization' of information processes.
People can now do things with computers and video appliances that,
less than a few generations ago, would have been inconceivable.

While the range of variations is steadily increasing, the contrasts
are diminishing. As the British historian of culture Alistair Laing
notes, electrification has led everywhere 'to the steady diminution of
social distinctions in the use of light, and to its general availability'.[30]

Wherever electric light was introduced, it was soon more or less taken for granted and no longer considered a luxury. The same can be said of numerous other facilities made possible by electricity, that have come to represent a level of comfort, hygiene and safety which in the industrialized world is regarded as normal. In the rich countries, deviations from this norm are to be found only in subcultures and slums. It has become almost standard practice to maintain both a constant level of artificial lighting and an even indoors temperature – achieved automatically, in cold winters by heating and in hot summers by cooling.

All these provisions require energy, produced by fuel which is sometimes brought in from far away. The continuous consumption of large quantities of fuel is causing side-effects that are, it is increasingly evident, affecting the whole of humanity. These side-effects include, on the one hand, increased emission of combustion gases into the air and, on the other, the imminent depletion of the earth's stock of fossil fuels. Awareness of the full extent of these costs has dawned only slowly, and even today experts disagree about the true measure of the ecological consequences of the progressive burning up of fossil fuels.

One thing, however, seems certain: estimating the consequences is a matter for experts. It lies beyond the comprehension of untrained laypeople who lack both the apparatus to collect the necessary information and the intellectual skills to assess it. The uneven distribution of knowledge can be seen as one of the consequences of the long-term process of specialization and organization that the American ethnologist and prehistorian Walter Hough said began with custody over the communal fire.[31]

No one has been able either to foresee or to plan this process in advance. Its course as a whole has been blind and unsteered. Yet it is the result of human intentions. Each innovation in the control of fire was accomplished because people deliberately tried to do something more with fire than they had been able to do till then. But they could not possibly foresee all the next steps that their successors would make. Nor did they always anticipate the increases in dependency that almost inevitably followed the increases in control.

It was precisely the lengthening and tightening of the chains of dependency that prompted people to contemplate the effects of their interference with their natural environment. The first books on that theme appeared in the second half of the nineteenth century. In the twentieth century the concern of a few has developed into a world-wide environmental movement. Increasing numbers of people are willing to acknowledge the possibility that the combination of intensive and extensive growth is engendering a rising fuel consumption that sooner or later will have disastrous consequences.[32]

There is a growing awareness that – unless, as a grim alternative, a major catastrophe occurs – for at least another few generations, further increases in world population are to be expected and, at the same time, rising demands for a higher standard of living. In these circumstances, it is becoming more and more difficult to ignore the problem of how the earth and the stratosphere will react, if and when the time comes that there are 10 billion people consuming as much fuel per person as the average citizen of the United States is consuming today.

To avert global disaster, two general strategies seem possible in principle: a reduction in the use of energy, and a switch to alternative energy sources, such as wind, water or nuclear power. The first strategy implies a change in the regulation of social relations and individual impulses; the second, a further extension of human control over 'extra-human' natural processes. As noted in Chapter 8, in some areas – for example, what we might call, almost paradoxically, the management of forest fires – a policy of 'controlled tempering of controls' is now becoming generally accepted. With regard to fuel consumption, this formula does not seem to be readily applicable, but even so there are plenty of attempts to curb unbridled growth. The fact that the different strategies are seriously discussed is in itself an indication of a new phase in the civilizing process.

Although there are some signs of a new sense of economy in the use of fuel, for the time being more effective results are expected from the second strategy – from tapping other sources of energy such as wind, water or nuclear power, which, it is hoped, can be exploited more efficiently and cause less pollution. As a process of technological innovation, this search is certainly guided by planning and co-ordination, but it is also motivated to a large extent by the uncontrolled pressure of competition – among states, among industrial companies and among scientific centres.

For several decades, the largest capital investments on this front were made in exploring the possibilities of nuclear fission. Confidence in these procedures has been severely shaken, however – first, by alarming theoretical calculations concerning the dangers of radioactive radiation, and then by accidents that have occurred in the United States and Western Europe, and most of all by the explosion in the nuclear reactor at Chernobyl in April 1986, which, according to the Russian physicist Zhores Medvedev, occasioned 'the most frightening catastrophe of modern industrial history'.[33]

Many experts point to nuclear fusion as a viable alternative to fission. The promise of nuclear fusion sounds almost too good to be true. The procedure is said to yield energy that will be both 'clean', causing no air pollution or radioactive fallout, and 'cheap', for the elements needed to bring about nuclear fusion are plentiful throughout the world. Once the breakthrough to profitable production were made, it would be possible to make this form of energy available everywhere in virtually limitless quantities.

As yet the promise has not been fulfilled. An announcement made in March 1989 that it was possible with relatively simple means to produce 'cold fusion' has turned out to be incorrect.[34] Efforts are now wholly concentrated upon 'hot fusion', which has to take place under conditions comparable to those at the centre of the sun. There, at a temperature of 15 million degrees, processes of nuclear fusion are occurring continuously; they form the ultimate source from which all energy in our solar system is derived. For controlled 'hot fusion' on our own planet, temperatures of at least 100 million degrees are required.

Technically such temperatures can already be achieved. At present, however, the procedure demands investments in energy which are so high as to be counter-productive. The financial expenditure required for exploring further possibilities is enormous, and this has induced the major competitors in the scientific race to combine their efforts. Since 1983 a consortium of Western European states including such non-EC members as Sweden and Switzerland has operated a laboratory at Culham near Oxford, called JET (Joint European Torus), where plasma temperatures of up to 140 million degrees have been achieved. In the late 1980s plans were launched for an even more all-encompassing international venture, called ITER (International Thermonuclear Experimental Reactor), under the auspices of the EC, Japan, the Soviet Union and the United States.[35]

For the time being, the ability to produce temperatures in the order of 150 million degrees may well deserve to be counted as a climax in the process of increasing human control of fire. No less striking is the degree of international cooperation in the movement towards harnessing nuclear fusion. In the Palaeolithic age, no group could continue to stay aloof from the human monopoly over the use of fire; similarly in our time, no state or nation will be able to close itself off from the contemporary developments in the fire regime at a global level.

The domestication of fire has made human life more comfortable and more complicated. The ubiquitous fires with their destructive potential and their never-ending need for fuel exert persistent pressures on society which have assumed different guises in successive stages. Because of the advances in specialization and organization, some of these demands are hardly felt by most people in contemporary industrial society. This does not mean, however, that they have disappeared. Every generation has to learn anew how to cope with fire. It does not have to master the same techniques as its predecessors, but its members do have to acquire the general capacity to live in a group-with-fire. In contemporary society this still means that all individuals have to obtain some elementary knowledge about fire itself, but, even more importantly, that they learn to participate in the social organization of the fire regime; and, it is to be hoped, that they gain some understanding of that regime.*

The trends noted in this book towards increasing use of fire, in a

---

* Specialist literature on fire regimes may be found in the library of the Fire Service College at Moreton-in-Marsh, Gloucestershire, Great Britain.

more concentrated form, under conditions of continuously advancing specialization and organization have contrived to make the control of fire seemingly more simple but actually far more complex. As a result of these trends, more and larger fires have been caused by humans in the twentieth century than in any previous age. People have derived more comfort from fire than ever before, and they have inflicted greater damage and suffering with it. At present, even more than in the past, the control of fire itself requires control; as such, it remains a central problem of human civilization.

NOTES

INTRODUCTION

1. Frazer 1930b; Lévi-Strauss 1969; 1972.
2. Cf. Bachelard 1964, pp. 59–82; Prigogine and Stengers 1984, pp. 103–209.
3. Darwin 1989, p. 49.
4. Tylor 1870, pp. 231–9.
5. Cf. Stewart 1956; Sauer 1952, pp. 10–18; Sauer 1981, pp. 129–56.
6. Elias 1978a, pp. 3–4.
7. Cf. Goudsblom 1980, pp. 51–74.
8. Renfrew 1972; 1976.
9. Cf. Goudsblom 1980, pp. 51–83. Whereas, in our modern languages, the noun 'civilization' is derived from a verb, 'culture' is not. We can say that people 'civilize' each other and themselves, but the same idea cannot be conveyed so neatly with the word culture.
10. Benedict 1935. See also pp. 15–16.
11. See, e.g., Festinger 1983; Hallpike 1986; Hillel 1991; Lenski *et al.* 1991; Stavrianos 1990.
12. Elias 1982, p. 276.
13. Elias 1978a, p. 160.
14. On the documentary value of this film see Liebermann 1982 and Perlès 1982.
15. McNeill 1976.
16. Elias 1978b, pp. 156–7. See also Goudsblom 1977, pp. 137–43.

CHAPTER 1

1. P. D. Moore 1982, p. 10.
2. Cf. Pyne 1982, pp. 10–11.
3. R. Brewer 1988, pp. 88–9.
4. Von den Steinen 1894, p. 220.
5. Von den Steinen 1894, p. 221.
6. Von den Steinen 1894, pp. 220–21.
7. Cf. H. T. Lewis 1972; 1989.
8. The Dutch zoologist Professor K. M. Voous drew my attention to this possibility. Claims

that birds do deliberately carry fire are made by Allaby 1982 and Burton 1959. See also Armstrong 1958, pp. 175–9; Bendell 1974.
9. Clark and Harris 1985.
10. See, e.g., Forni 1984.
11. Dart 1948.
12. Cf. Oakley 1955; Gowlett *et al.* 1981; 1982; Isaac 1982. For a recent survey see James 1989 and the discussion following it in the same issue of *Current Anthropology.* See also Renfrew and Bahn 1991, p. 226.
13. See Perlès 1977, pp. 13–26.
14. On the distinction between chronology and phaseology see Goudsblom, Jones and Mennell 1989, pp. 11–26.
15. Perlès 1977, p. 30, and Perlès 1981.
16. Freud 1961, p. 90, and Freud 1964.
17. McGrew 1989; Brink 1957. See also Chapter 2, pp. 24–5.
18. Cf. Elias 1982, pp. 91–115.
19. Cf. Morris 1967; Kortlandt and Kooij 1963; Kortlandt 1972.
20. Washburn and Lancaster 1968, p. 221.

CHAPTER 2

1. Brink 1957, p. 247.
2. Brain 1981, p. 273. See also Brain and Sillern 1988.
3. Chatwin 1987, pp. 260–92.
4. Howell 1965, p. 84. See also Howell 1966.
5. Konner 1982, p. 51. See also Johanson and Edey 1982, p. 73; Johanson and Shreeve 1989, p. 221.
6. Binford 1989, pp. 383–422. See also p. 473.
7. Cf. Stewart 1956, pp. 119–20.
8. Goodale 1971, p. 169; Lee 1979, p. 234; Shostak 1981, p. 101.
9. Boserup 1970, p. 15.
10. Stewart 1956, p. 118.
11. Cf. Shostak 1981, p. 11. See also Chapter 8, pp. 189–90.
12. Sauer 1981, p. 340. See also Talbot 1989, pp. 18–19.
13. Cf. Roberts 1989, pp. 42–61.
14. Cf. Edwards 1988.
15. Cronon 1983, p. 51. The earlier quotations are from Thomas Morton and Roger Williams as cited by Cronon 1983, pp. 49–51. See also Day 1953; Pyne 1982; Russell 1983.
16. Pyne 1982, p. 84.
17. Hallam 1975, pp. 16–28. See also Blainey 1975a, pp. 67–83.
18. For the 'burning controversy' on this issue see R. Jones 1969; Horton 1982; R. Jones 1989.
19. Scott Nind (1831) and J. L. Stokes (1846) as quoted by Hallam 1975, pp. 32–3.
20. Hallam 1975, pp. 105–7. See also H. T. Lewis 1989.
21. S. Brewer 1978, p. 191.
22. Stahl 1984. See also Peters and O'Brien 1984.

23. Perlès 1979, pp. 7–9.
24. Cf. Leibowitz 1985, pp. 64–7.
25. Cf. Perlès 1977, p. 101.
26. Stone 1979, p. 30.
27. Cf. Müller 1972, p. 4.
28. Cf. Goudsblom 1977, pp. 175–80.
29. On the concept of survival units see Elias 1978b, pp. 138–9.
30. Cf. Jones in Goudsblom, Jones and Mennell 1989, pp. 27–62.
31. Perlès 1981.
32. Cf. Baker 1984.
33. McNeill 1976, pp. 27–8.
34. Hudson 1976, pp. 235–6.
35. Lévi-Strauss 1976. See also Klaatsch 1920, p. 100; Lumsden and Wilson 1983, p. 100.
36. Cf. Trigger 1976, pp. 73–4.
37. Cf. Hough 1916.
38. See, e.g., Hudson 1976; Staal 1983.
39. Bachelard 1964, p. 22.
40. See also Chapters 8 and 9, pp. 191, 197–8.

CHAPTER 3

1. Gowlett 1984, pp. 10–11. Cf. Murdock 1968.
2. Cf. Clutton-Brock 1987, pp. 9–16.
3. Cf. McNeill 1989.
4. Cf. Cohen 1977.
5. H. T. Lewis 1972.
6. Mellars 1976.
7. Forni 1984.
8. R. Jones 1969; Hallam 1975; Blainey 1975a, pp. 67–83; Horton 1982. See also Cumberland and Whitelaw 1970; Henley 1982.
9. Cf. Conklin 1961.
10. See also Chapter 8, pp. 188–93.
11. Dickens, *American Notes*, 1842, p. 165. Quoted by J. G. D. Clark 1952, p. 93.
12. Allan 1965, p. 67. See also De Schlippe 1956.
13. Steensberg 1979; 1980.
14. Geertz 1966, p. 26.
15. J. G. D. Clark 1952, p. 92.
16. Sahlins 1972, p. 35.
17. J. G. D. Clark 1952, p. 98.
18. Cf. Sigaut 1975, p. 283. See also Kuhnholtz-Lordat 1938; Raumolin 1987.
19. Rowley-Conwy 1981. See also Barker 1985.
20. Cf. Carneiro 1961; Mann 1986, pp. 73–104. See also Simmons 1989, p. 169; Wolf 1959, pp. 58–62.
21. Boserup 1965, p. 30.

CHAPTER 4

1. Sahlins 1972, pp. 1–40; Harris 1977, pp. 9–13.
2. Colgrave and Mynors 1969, pp. 183–4.
3. Cf. Boyce 1979; Duchesne-Guillemin 1962; Mokri 1982; Staal 1983.
4. Gellner 1988, p. 154.
5. Cf. Goudsblom, Jones and Mennell 1989, pp. 79–92.
6. Renfrew and Bahn 1991, p. 292.
7. Muhly 1988, p. 7.
8. Muhly 1988, p. 2.
9. Muhly 1988, p. 5.
10. In Clarke, Cowie and Foxon 1985, p. 178. See also Coghlan 1975, pp. 50–74; Wertime and Wertime 1982.
11. Renfrew 1972, p. 320.
12. Renfrew 1972, p. 339.
13. Lenski *et al.* 1991, p. 168.
14. Cf. Eliade 1962, p. 85. See also Eliade 1964, pp. 470–74.
15. Gurney 1990, p. 95. The word omitted is 'legitimate'.
16. Ebeling 1957, p. 56.
17. Pritchard 1969, p. 167 (n. 25).
18. Pritchard 1969, p. 176 (ns. 226–7).
19. Cf. Lloyd 1984, p. 44. See also Goudsblom, Jones and Mennell 1989, pp. 67–8.
20. McGrew 1991, p. 13.
21. Pritchard 1969, p. 209 (the italics are mine).
22. Neufeld 1951, pp. 30–31. See also Hoffner 1963, pp. 75–6, 240–41.
23. Bankoff and Winter 1979, p. 35.
24. Braudel 1981, p. 267.

CHAPTER 5

1. Cf. Gottwald 1979, pp. 25–31; Lemche 1985, pp. 357–85; Lemche 1988, pp. 29–73; Miller, Maxwell and Hayes 1986, pp. 54–79; Rogerson and Davies 1989, pp. 346–75.
2. For other interpretations of Abraham's sacrifice see Spiegel 1969. See also Morgenstern 1963.
3. All biblical quotations are from the Authorized King James Version.
4. Cf. Kranendonk 1990, pp. 86–91. See also Chapter 8, pp. 179–80.
5. Cf. Goudsblom 1989.
6. Harris 1977, pp. 117–19.
7. On attitudes to magic in ancient Israel see M. Weber 1952, pp. 219–25 and Zeitlin 1984, pp. 30–32. I suspect that Zeitlin underestimates its importance.
8. Rogerson and Davies 1989, pp. 118, 132.
9. Elias 1978a, pp. 76–80.
10. Cf. Gottwald 1979; Lemche 1988, pp. 75–117; Miller, Maxwell and Hayes 1986, pp. 74–9.

11. Naveh 1974, p. 408.

12. Cf. Hanson 1983.

13. Quoted from the translation by Jonathan A. Goldstein, Doubleday & Co., Garden City, N Y, 1983.

CHAPTER 6

1. Cf. Rogerson and Davies 1989, p. 72.

2. Cf. Finley 1977, pp. 142–58; Renfrew 1972, pp. 68–9; Snodgrass 1974.

3. Pliny, *Natural History* 36.10.

4. See, e.g., Edsman 1949; Simons 1949; Furley 1981; Burkert 1985, pp. 60–64. Scholars sometimes fail to distinguish between actually existing conditions and phantoms which people only professed to believe in. Thus Burkert begins his discussion of fire rituals as follows: 'Fire is one of the foundations of civilized life. It is the most primitive protection from beasts of prey, and so also from evil spirits.' In this statement 'emic' and 'etic' categories are mixed, relieving the author and his readers of the task of making a clear distinction between the real world in which people lived and the world of their imagination.

5. Cf. Finley 1977. Unfortunately the military functions of the *oikos* do not get as much attention in Finley's reconstruction as the economic functions.

6. All extracts from the *Iliad* are from the translation by Martin Hammond, Penguin Books, Harmondsworth, 1987.

7. Graz 1965, pp. 345–50.

8. Brain 1981, pp. 97–8.

9. Burkert 1985, p. 191.

10. Quoted by Ste Croix 1981, p. 10. See also p. 210.

11. All extracts from Hesiod are from the translation by Dorothea Wender, Penguin Books, Harmondsworth, 1973. See also West 1978, pp. 336–7 for further comment on this passage, and pp. 1–91 for the general background. See also Zimmermann 1947, pp. 226–53.

12. Cf. M. Weber 1930.

13. Cf. Ste Croix 1981, p. 328.

14. The extract from Xenophon is from the translation by Hugh Tredennick and Robin Waterfield, Penguin Books, Harmondsworth, 1990. The only reference I have found in the *Georgics* is: 'Learn, too, to burn in your stalls fragrant cedar and with fumes of Syrian gum to banish the noisome water-snakes.' (III, 414–15)

15. All extracts from Herodotus are from the translation by Aubrey de Sélincourt, Penguin Books, Harmondsworth, 1972.

16. The italics are mine. All extracts from Thucydides are from the translation by Rex Warner, Penguin Books, Harmondsworth, 1972.

17. Cf. Meiggs 1982, p. 375.

18. Cf. Ferrill 1985, pp. 204–6.

19. Hanson 1983, pp. 11–36.

20. Van Creveld 1991, p. 28.

21. Cf. Hanson 1983, pp. 59–60, 92.

22. White 1984, p. 44.

23. Burford 1972, p. 73.

24. Burford 1972, p. 72. Cf. Forbes 1964, pp. 81–2; Graves 1955 I, p. 88.

25. Burford 1972, p. 122.

26. Cf. Burford 1972, pp. 122–3. See also Forbes 1958b, p. 74.

27. Burford 1972, pp. 178, 211.

28. Fustel de Coulanges 1956, p. 25.

29. Carcopino 1941, p. 47.

30. Carcopino 1941, pp. 47–8.

31. Cf. Freud 1964b, p. 102.

32. Werner 1906, pp. 9–47.

33. Cf. Hopkins 1978, pp. 96–8.

34. Vitruvius, *The Ten Books on Architecture* 2:1. See also Yavetz 1958.

35. All extracts from Juvenal are from the translation by Peter Green, Penguin Books, Harmondsworth, 1967.

36. Werner 1906, pp. 48–50.

37. Cf. Frier 1980, pp. 21–30, 63, 75; Hermansen 1981, pp. 207–25.

38. Plutarch's *Lives*, 'Crassus', section 2. See also Frier 1980, pp. 32–4.

39. Reynolds 1926, pp. 19–21.

40. Rainbird 1986, pp. 150–51. See also Reynolds 1926; Robinson 1977.

41. Newbold 1974, p. 858.

42. Cf. Ste Croix 1981, pp. 13–14, 208–26.

43. Newbold 1974, p. 861.

44. Cf. Tacitus, *The Annals* 4:64, 6:45.

45. These extracts from the Younger Pliny are from the translation by Betty Radice, Penguin Books, Harmondsworth, 1963.

46. See, e.g., Frisch 1963; Bradbury 1954.

47. Cf. Frier 1980, pp. 142–7.

48. Cf. Ste Croix 1981, pp. 320–21; MacMullen 1974, p. 66. See also Hopkins 1983, p. 29 (n. 39).

49. Ste Croix 1981, p. 498.

50. Muir 1985, p. 194.

51. Fustel de Coulanges 1956, p. 26.

52. Cf. Adkins 1960.

53. Simons 1949.

54. For a different interpretation and further references see Beard 1980.

55. Forbes 1958b, pp. 180–82.

56. All extracts from Pausanias are from the translation by Peter Levi, Penguin Books, Harmondsworth, 1971. See also Furley 1981, pp. 116–51.

57. This extract from Tacitus is from the translation by John Jackson, William Heinemann, London, 1949.

58. Hughes and Thirgood 1982, p. 196.

59. Plato, *Phaedo* 110e.

60. Meiggs 1982, p. 377.

61. Hughes and Thirgood 1982, p. 207.

CHAPTER 7

1. Cf. Stone and Mennell 1980, pp. 25–41.
2. Herrin 1987, pp. 295–306.
3. Cf. Boyce 1979; Mokri 1982; Staal 1983.
4. Cf. Freud 1955.
5. Cf. Hagger 1991, pp. 88–141.
6. Frazer 1930a I, p. 328.
7. Frazer 1930a I, p. vii.
8. Frazer 1930a II, p. 39.
9. Elias 1978a, pp. 203–4. See also Darnton 1984, pp. 87–8.
10. Freudenthal 1931, pp. 301–10.
11. Cf. Hobsbawm and Ranger 1983.
12. R. I. Moore 1987, pp. 13–15.
13. Sumption 1978, pp. 227–30.
14. Cf. Le Goff 1984.
15. Quoted by Camporesi 1990, p. 81.
16. Goudsblom 1980, pp. xii–xiv.
17. Quoted by Ellis Davidson 1973, p. 66.
18. Cf. Forbes 1959, pp. 70–90; Partington 1960, pp. 10–41; Finó 1970; Ellis Davidson 1973; Haldon and Byrne 1977.
19. Van Creveld 1991, p. 15.
20. Haldon and Byrne 1977, p. 99.
21. Ellis Davidson 1973, pp. 65–6.
22. McNeill 1982, p. 87.
23. Needham 1985, pp. 6–14.
24. McNeill 1982, p. 39.
25. Oman 1926, p. 226.
26. McNeill 1982, pp. 95–9.
27. Cf. Perrin 1979.
28. Gutmann 1980, p. 63.
29. McNeill 1982, p. 100. See also Cipolla 1965, pp. 137–40.
30. Quoted by Needham 1985, p. 2.
31. Cf. Hemming 1983, pp. 110–17. See also McNeill 1976, pp. 1–5; Collins 1986, pp. 85–92.
32. See also McNeill 1982, pp. 79–116.
33. Cipolla 1972, pp. 18, 71, 80.
34. See, e.g., Bowsky 1981, pp. 296–7. In his book on medieval Siena Bowsky treats the organization of the fire brigade only in passing. Most historians pay even less attention to the subject. It is usually regarded as a topic of local interest only and is not discussed at all in the general works on urban history by Sjoberg 1960, and Hohenberg and Lees 1985.
35. Green-Hughes 1979, pp. 14–15.
36. Braudel 1981, pp. 266–8.
37. Meyer and van den Elzen 1982, p. 98.
38. Meyer and van den Elzen 1982, p. 11.

39. Lane 1973, pp. 439–40.
40. Meyer and van den Elzen 1982, pp. 12–13.
41. Latham and Matthews 1972, pp. 267–82. See also Bell 1920; Milne 1986.
42. McCloy 1946, pp. 86–9.
43. Vries 1984, pp. 88–9.
44. Frost and Jones 1989, p. 333.
45. Jones, Porter and Turner 1984.
46. E. L. Jones 1987, pp. 33–4. See also Braudel 1981, pp. 268–73.
47. Cf. Kitching 1981.
48. Cf. Dorwart 1971, pp. 293–304; McCloy 1946, pp. 99–105.
49. Gales and van Gerwen 1988, pp. 61–2.
50. Clayton 1971, pp. 34–48.
51. Dorwart 1971, pp. 299–302.
52. Cf. Konvitz 1985, p. 124.
53. Brimblecombe 1987, p. 16.
54. Te Brake 1975.
55. Zeeuw 1978. See also Unger 1984.
56. Wrigley 1988, p. 55.
57. Cf. Perlin 1989, pp. 163–245; K. Thomas 1983, pp. 192–223. See also Finberg 1976, p. 74.
58. Slicher van Bath 1963. See also Simmons 1989, p. 166.
59. Braudel 1981, p. 299. It is an intriguing problem why it has taken so long for a system of central heating similar to that used in ancient Rome to be introduced in modern Europe. We may well wonder about the extent to which status has played a part in making people increasingly sensitive to cold.
60. E. Weber 1976, p. 17.
61. Franklin 1986, p. 213.
62. Cf. K. Thomas 1971, p. 18.
63. *Njall's Saga*, which is situated in Iceland around 1200, contains an episode of warrior-farmers killing one of their enemies and his family by burning down his house. This way of using fire to eliminate an enemy whom one was unable to defeat in combat was regarded as dishonourable. Cf. Magnusson and Pálsson 1960; Byock 1988.
64. Radzinowicz 1948, p. 9.
65. Durkheim 1951.
66. Cf. Abbiateci 1970; Wieërs 1986.
67. Willems 1981.
68. Cf. Swaan 1988, pp. 13–51.
69. Gabriel Tarde, *Essais et mélanges sociologiques* (Lyons, 1895), p. 121. Quoted by E. Weber 1976, p. 16.
70. Claverie and Lamaison 1982, pp. 17–23.
71. D. Jones 1982, pp. 34, 35. See also Hobsbawm and Rudé 1969.
72. Price 1983, pp. 392–5; Wright 1983, pp. 156–62; Schulte 1984.
73. Cf. Braudel 1981, pp. 399–402. On Agricola, see Hoover and Henry 1950.
74. Cf. Forbes 1970, pp. 81–2, 174–5, 272. For the oldest of the three temperature scales (devised by Fahrenheit in 1714) the point of departure was still the warmth of the human body.

75. Boerhaave, *Eléments de Chimie*, 2 vols., Leyden, 1752, I, p. 144. As quoted by Bachelard 1964, p. 60.

76. Quoted by Prigogine and Stengers 1984, p. 55.

77. Among those who struggled with the problem of weighing the effects of fire were Voltaire and Mme de Châtelet. See the elegant description in Forster 1936, pp. 199–204.

CHAPTER 8

1. J. C. D. Clark 1985, p. 66.

2. Cf. E. L. Jones 1988, pp. 13–27; Mokyr 1990, pp. 81–4; Wallerstein 1989, pp. 3–33; Wrigley 1987, pp. 2–4; Wrigley 1988, pp. 8–12.

3. Wrigley 1988.

4. Mathias 1983, p. 221. For figures on intensive growth in Great Britain see p. 222.

5. McEvedy and Jones 1978, p. 349.

6. Simmons 1989, p. 379. On the measurement of energy, see also Foley 1987, pp. 44–54.

7. For their impressions see Clayre 1977, pp. 117–31. See also Trinder 1982.

8. Quack 1915, p. 131.

9. Thomas S. Ashton (1948) and Paul Mantoux (1946) as quoted by Sicilia 1986, p. 287; Forbes 1958a, p. 150.

10. Cf. Hoskins 1970, p. 215.

11. Von Tunzelmann 1978, pp. 286–7.

12. Cf. Blainey 1975b, pp. 116–22.

13. Elias 1982, pp. 149–61.

14. Mathias 1983, p. 138.

15. Quoted in Clayre 1977, pp. 118–19.

16. Beaver 1985, pp. 18–28.

17. Briggs 1988, p. 181; Beaver 1985, p. 19.

18. Briggs 1988, p. 195.

19. Simmons 1989, pp. 239–43.

20. Cf. Williams 1982, pp. 64–79.

21. Gleichmann 1983. See also Melbin 1987.

22. Cf. Yergin 1991, pp. 588–652.

23. Lyons 1985, p. 111. See also Rosen 1986.

24. Frost and Jones 1989, pp. 334–5.

25. E. L. Jones 1987, pp. 33–4.

26. Frost and Jones 1989, p. 341.

27. The following account is based on Douwes 1968.

28. Many examples of this ethos are to be found in the literature on the history of fire brigades in various countries. See, e.g., Wallington 1989.

29. Green-Hughes 1979, p. 34.

30. Lyons 1985, pp. 2–4; Kolata 1987, p. 281.

31. Kolata 1987, p. 281.

32. Kolata 1987, p. 282.

33. Lyons 1985, pp. 140–42.

34. Mumford 1961, pp. 632–8.

35. Longmate 1983, p. 221.

36. Serious estimates of the number of victims of the air raid on Dresden vary from 35,000 (Beck 1986, pp. 177–80) to 135,000 (Longmate 1983, p. 341). These enormous discrepancies regarding an event in 1945 must raise doubts about the reliability of figures relating to the numbers of victims of armed conflict and other disasters in the more remote past.

37. Salomon 1931, p. 131.

38. Churchill 1930, p. 162.

39. Multatuli 1982, p. 276.

40. Cf. Harden 1980, p. 167.

41. Malinowksi 1967, pp. 11–12.

42. Cf. De Schlippe 1956; Christiansen 1981; Raumolin 1987.

43. Conklin 1957.

44. Carneiro 1961.

45. Myers 1984, p. 150.

46. Kirby 1987, p. 14.

47. Kirby 1987, p. 19. For similar statements about Neolithic Scandinavians see Hodder 1990, pp. 199–200.

48. Blainey 1975a, p. 76.

49. Korem 1985, p. 24.

50. Schoffeleers 1978, pp. 3–4; Schoffeleers 1971; Chapman and White 1970, pp. 31–4.

51. Kirby 1987, p. 18.

52. Many examples are given by Pyne 1982.

53. H. T. Lewis 1989, p. 940.

54. Harold H. Biswell in Kozlowski and Ahlgren 1974, p. 356.

55. A. J. Kayll in Kozlowski and Ahlgren 1974, p. 503. Brief quotations are taken from other contributions, on pp. 170, 180, 182, 247, 271, 281, 435.

56. Cf. Elias and Dunning 1986, pp. 44–9.

CHAPTER 9

1. Personal communication from the late Dick Hillenius. See also Komarek 1967, pp. 151–5.

2. Festinger 1983, pp. 16–18.

3. Bachelard 1964, p. 11.

4. Elias 1978a, p. 82.

5. Goodale 1971, p. 34.

6. Goodale 1971, p. 36.

7. Shostak 1981, p. 107.

8. Marshall 1976, p. 291.

9. Cf. Ney and Gale 1989.

10. Cf. Lyons 1985, pp. 136–57.

11. Kafry 1990, p. 60.

12. Kafry 1990, p. 54.

13. Grinstein 1952, p. 418.

14. Joseph 1960, p. 102.

15. Vreeland and Levin 1990, pp. 40–41.

16. Lewis and Yarnell 1951, p. v (the italics are mine).

17. See, e.g., Canetti 1946; Du Maurier 1938. On films, see Armstrong and Armstrong 1990, pp. 128–30; on art, see Draxler 1987; on religion, see Hagger 1991.

18. Durkheim 1951.

19. Ribeiro 1968, pp. 3–4.

20. Eliade 1962, p. 144 (n.).

21. Radcliffe-Brown 1922, p. 472.

22. Laing 1982, pp. 41–2.

23. Cf. Bourdieu 1984.

24. Cf. Hudson 1976, pp. 255–7.

25. Rushdie 1991, p. 139.

26. Simmons 1989, p. 214.

27. 'Frank', as quoted by Angela Johnson in the *Independent*, 3 April 1991.

28. Thornton 1984, p. 15. See also O'Dea 1958, pp. 1–26; Laing 1982.

29. Elias 1982, pp. 251–8.

30. Laing 1982, pp. 6–7.

31. Hough 1926, p. 165.

32. See, e.g., Bramwell 1983; Hardin 1985; Mitchell 1991.

33. Medvedev 1990, p. ix.

34. Cf. Close 1990; Mallove 1991.

35. Cf. Maple 1987.

BIBLIOGRAPHY

Abbiateci, André, 'Les incendiaires dans la France du XVIIIe siècle. Essai de typologie criminelle', *Annales. Economies, Sociétés, Civilizations* 25 (1970), pp. 229–48.

Adkins, W. H., *Merit and Responsibility: A Study in Greek Values*, Oxford University Press, 1960.

Allaby, Michael, *Animal Artisans*, Knopf, New York, 1982.

Allan, William, *The African Husbandsman*, Oliver & Boyd, Edinburgh, 1965.

Armstrong, Edward A., *The Folklore of Birds: An Enquiry into the Origin and Distribution of Some Magico-Religious Traditions*, Collins, London, 1958.

Armstrong, Richard B. and Mary Willems Armstrong, *The Movie List Book: A Reference Guide to Film Themes, Settings, and Series*, McFarland, Jefferson, NC, 1990.

Bachelard, Gaston, *The Psychoanalysis of Fire* (translated from the French, 1938), Beacon Press, Boston, 1964.

Baker, Paul T., 'The Adaptive Limits of Human Populations', *Man* (n.s.) 19 (1984), pp. 1–14.

Bankoff, H. Arthur and Fredrick A. Winter, 'A House-Burning in Serbia', *Archaeology* 32, 5 (Sept. 1979), pp. 8–14.

Barker, Graeme, *Prehistoric Farming in Europe*, Cambridge University Press, 1985.

Beard. M., 'The Sexual Status of the Vestal Virgins', *Journal of Roman Studies* 70 (1980), pp. 12–27.

Beaver, Patrick, *The Match Makers*, Henry Mellard, London, 1985.

Beck, Earl R., *Under the Bombs: The German Home Front 1942–1945*, University Press of Kentucky, Lexington, 1986.

Bell, Walter George, *The Great Fire of London in 1666*, John Lane, London, 1920.

Bendell, J. F., 'Effects of Fire on Birds and Mammals', in Kozlowski and Ahlgren 1974, pp. 73–138.

Benedict, Ruth, *Patterns of Culture*. Routledge and Kegan Paul, London, 1935.

Binford, Lewis R., *Debating Archaeology*, Academic Press, San Diego, 1989.

Birks, Hilary H., H. J. B. Birks, Peter Emil Kaland and Dagfinn Moe (eds.), *The Cultural Landscape – Past, Present and Future*, Cambridge University Press, 1988.

Blainey, Geoffrey, *Triumph of the Nomads: A History of Ancient Australia*, Macmillan, Melbourne, 1975a.

The Tyranny of Distance: How Distance Shaped Australia's History (2nd edn), Macmillan, Melbourne, 1975b.

Boserup, Ester, *The Conditions of Agricultural Growth*, Aldine, Chicago, 1965.
   Woman's Role in Economic Development, Allen and Unwin, London, 1970.
Bottéro, Jean, 'Notes sur le feu dans les textes Mésopotamiens', in *Le Feu dans le Proche-Orient
   antique. Aspects linguistiques, archéologiques, littéraires*, Actes du Colloque de Strasbourg (9
   and 10 June 1972), Brill, Leiden, 1973, pp. 9–30.
Bourdieu, Pierre, *Distinction. A Social Critique of the Judgement of Taste* (translated from the
   French, 1979), Routledge & Kegan Paul, London, 1985 and 1990.
Bowsky, William M., *A Medieval Italian Commune: Siena under the Nine, 1287–1355*,
   University of California Press, Berkeley, 1981.
Boyce, Mary, *Zoroastrians: Their Religious Beliefs and Practices*, Routledge & Kegan Paul,
   London, 1979.
Bradbury, Ray, *Fahrenheit 451*, Hart-Davis, New York, 1954.
Brain, C. K., *The Hunters or the Hunted? An Introduction to African Cave Taphonomy*, University
   of Chicago Press, 1981.
Brain, C. K., and A. Sillen, 'Evidence from the Swartkrans cave for the earliest use of fire',
   Nature 336 (Dec. 1988), pp. 464–6.
Bramwell, Anne, *Ecology in the Twentieth Century: A History*, Yale University Press, New
   Haven, 1989.
Braudel, Fernand, *Civilization and Capitalism 15th–18th Century. 1. The Structures of Everyday
   Life* (translated from the French, 1979), Fontana, London, 1981.
Brewer, Richard, *The Science of Ecology*, Saunders, Philadelphia, 1988.
Brewer, Stella, *The Forest Dwellers*, Collins, London, 1978.
Briggs, Asa, *Victorian Things*, Penguin Books, Harmondsworth, 1988.
Brimblecombe, Peter, *The Big Smoke: A History of Air Pollution in London since Medieval
   Times*, Methuen, London, 1987.
Brink, A. S., 'The Spontaneous Fire-controlling Reactions of Two Chimpanzee Smoking
   Addicts', *South African Journal of Science* 53 (1957), pp. 241–7.
Burford, Alison, *Craftsmen in Greek and Roman Society*, Thames & Hudson, London, 1972.
Burkert, Walter, *Greek Religion: Archaic and Classical* (translated from the German, 1977),
   Basil Blackwell, Oxford, 1985.
Burton, Maurice, *Phoenix Re-born*, Hutchinson, London, 1959.
Byock, Jesse L., *Medieval Iceland: Society, Sagas, and Power*, University of California Press,
   Berkeley, 1988.
Camporesi, Piero, *The Fear of Hell: Images of Damnation and Salvation in Early Modern Europe*
   (translated from the Italian, 1987), Polity Press, Cambridge, 1990.
Canetti, Elias, *Auto da fé* (translated from the German, 1935), Jonathan Cape, London,
   1946.
Canter, David (ed.), *Fires and Human Behaviour* (2nd edn), David Fulton, London, 1990.
Carcopino, Jérome, *Daily Life in Ancient Rome* (translated from the French, 1939), E. O.
   Lorimer, London, 1941.
Carneiro, Robert L., 'Slash-and-Burn Cultivation among the Kuikuru and Its Implications
   for Cultural Development in the Amazon Basin', in Jonathan Wilbert (ed.), *The
   Evolution of Horticultural Systems in Native South America: Causes and Consequences*,
   Editorial Sucre, Caracas, 1961, pp. 47–67.
Chapman, J. D. and F. White, *The Evergreen Forests of Malawi*, Commonwealth Forestry
   Institute, Oxford, 1970.

Chatwin, Bruce, *The Songlines*, Jonathan Cape, London, 1987.

Christiansen, S., 'Shifting Cultivation – Survey of Recent Views', *Folk* 23 (1981), pp. 177–84.

Churchill, Winston S., *My Early Life*, Thornton Butterwort, London, 1930.

Cipolla, Carlo M., *Guns and Sails in the Early Phase of European Expansion 1400–1700*, Pantheon Books, London, 1965.

(ed.), *The Fontana Economic History of Europe. 1. The Middle Ages*, Fontana, Glasgow, 1972.

Clark, J. C. D., *English Society 1688–1832*, Cambridge University Press, 1985.

Clark, J. D. and J. W. K. Harris, 'Fire and Its Roles in Early Hominid Lifeways', *The African Archaeological Review* 3 (1985), pp. 3–27.

Clark, J. G. D., *Prehistoric Europe: The Economic Basis*, Methuen, London, 1952.

Clarke, D.V., T. G. Cowie and A. Foxon, *Symbols of Power at the Time of Stonehenge*, National Museum of Antiquities of Scotland, Edinburgh, 1985.

Claverie, Elisabeth and Pierre Lamaison, *L'impossible mariage: Violence et parenté en Gévaudan 17e, 18e et 19e siècles*, Hachette, Paris, 1982.

Clayre, Alasdair (ed.), *Nature and Industrialization*, Oxford University Press, 1977.

Clayton, G., *British Insurance*, Elek Books, London, 1971.

Close, Frank, *Too Hot to Handle: The Story of the Race for Cold Fusion*, W. H. Allen, London, 1990.

Clutton-Brock, Juliet, *A Natural History of Domesticated Animals*, Cambridge University Press, 1987.

Coghlan, H. H., *Notes on the Prehistoric Metallurgy of Copper and Bronze in the Old World* (2nd edn), Oxford University Press, 1975.

Cohen, Mark Nathan, *The Food Crisis in Prehistory: Overpopulation and the Origins of Agriculture*, Yale University Press, New Haven, 1977.

Colgrave, Bertram and R. A. B. Mynors (eds.), *Bede's Ecclesiastical History of the English People*, Clarendon Press, Oxford, 1969.

Collins, Randall, *Weberian Sociological Theory*, Cambridge University Press, 1986.

Conklin, Harold C., *Hanunóo Agriculture: A Report on an Integral System of Shifting Cultivation in the Philippines*, Food and Agricultural Organization, Rome, 1957.

'The Study of Shifting Cultivation', *Current Anthropology* 2 (1961), pp. 27–64.

Cronon, William, *Changes in the Land: Indians, Colonists, and the Ecology of New England*, Hill & Wang, New York, 1983.

Cross, M. F., 'A History of the Match Industry', *Journal of Chemical Education* 18 (1941), pp. 116–20, 277–82, 316–19, 380–84, 421–31.

Cumberland, Kenneth B. and James S. Whitelaw, *New Zealand*. The World's Landscapes 5, Longman, London, 1970.

Darnton, Robert, *The Great Cat Massacre and Other Episodes in French History*, Basic Books, New York, 1984.

Dart, R. A., 'The Makapansgat Proto-human Australopithecus prometheus', *American Journal of Physical Anthropology* (n.s.) 6 (1948), pp. 259–84.

Darwin, Charles, 'The Descent of Man and Selection in Relation to Sex' (2nd edn, 1877), in Paul H. Barrett and R. B. Freeman (eds.), *The Works of Charles Darwin*, vols. 21–2, William Pickering, London, 1989.

Day, Gordon M., 'The Indian as an Ecological Factor in the Northeastern Forest', *Ecology* 34 (1953), pp. 329–46.

De Schlippe, Pierre, *Shifting Cultivation in Africa: The Zande System of Agriculture*, Routledge & Kegan Paul, London, 1956.

Dorwart, Reinhold A., *The Prussian Welfare State Before 1740*, Harvard University Press, Cambridge, Mass., 1971.

Douwes, F. G. M., 'De grote brand in 1858', *Ons Amsterdam* 20 (1968), pp. 267–71.

Draxler, Helmut, 'Das brennende Bild: Eine Kunstgeschichte des Feuers in der neueren Zeit', *Kunstforum* 87 (1987), pp. 70–228.

Duchesne-Guillemin, J., 'Fire in Iran and in Greece', *East and West* 13 (1962), pp. 198–206.

Du Maurier, Daphne, *Rebecca*, Gollancz, London, 1938.

Durkheim, Emile, *Suicide: A Study of Sociology* (translated from the French, 1897), Free Press, Glencoe, Ill., 1951.

Ebeling, E., 'Feuerbekämpfung', in *Reallexikon der Assyriologie*, Vol. 3 (1957), p. 56.

Edsman, Carl-Martin, *Ignis divinus: Le feu comme moyen de rajeunissement et d'immortalité*, Gleerup, Lund, 1949.

Edwards, Kevin J., 'The Hunter-Gatherer/Agricultural Transition and the Pollen Record in the British Isles', in H. H. Birks *et al.*, 1988, pp. 255–66.

Eliade, Mircea, *The Forge and the Crucible* (translated from the French, 1956), Rider & Co., London, 1962.
   Shamanism: Archaic Techniques of Ecstasy (translated from the French, 1951), Pantheon, New York, 1964.

Elias, Norbert, *The Civilizing Process. 1. The History of Manners* (translated from the German, 1939), Urizen Books, New York, 1978a.
   What Is Sociology? (translated from the German, 1970), Hutchinson, London, 1978b.
   The Civilizing Process. 2. State Formation and Civilization (translated from the German, 1939), Basil Blackwell, Oxford, 1982.

Elias, Norbert and Eric Dunning, *Quest for Excitement: Sport and Leisure in the Civilizing Process*, Basil Blackwell, Oxford, 1986.

Ellis Davidson, H. R., 'The Secret Weapon of Byzantium', *Byzantinische Zeitschrift* 66 (1973), pp. 61–74.

Ferrill, Arther, *The Origins of War*, Thames & Hudson, London, 1985.

Festinger, Leon, *The Human Legacy*, Columbia University Press, New York, 1983.

Finberg, H. P. R., *The Formation of England 550–1042*, Paladin Books, London, 1976.

Finley, M. I., *The World of Odysseus* (2nd edn), Chatto & Windus, London, 1977.

Finó, J.-F., 'Le feu et ses usages militaires', *Gladius* 9 (1970), pp. 15–30.

Foley, Gerald, *The Energy Question* (3rd edn), Penguin Books, Harmondsworth, 1987.

Forbes, R. J., 'Power to 1850', in Charles Singer *et al.* (eds.), *A History of Technology*, Vol. 4, Clarendon Press, Oxford, 1958a, pp. 148–67.
   Studies in Ancient Technology: 6, Brill, Leiden, 1958b.
   More Studies in the Early Petroleum Industry, Brill, Leiden, 1959.
   Studies in Ancient Technology: 8, Brill, Leiden, 1964.
   A Short History of the Art of Distillation, Brill, Leiden, 1970.

Forni, G., 'From Pyrophytic to Domesticated Plants: The Palaeontological-linguistic Evidence for a Unitary Theory on the Origin of Plant and Animal Domestication', in W. van Zeist and W. A. Casparie (eds.), *Plants and Ancient Man: Studies in Palaeoethnobotany*, Balkema, Rotterdam, 1984, pp. 131–9.

Forster, E. M., *Abinger Harvest*, Edward Arnold, London, 1936.

Franklin, Benjamin, *The Autobiography and Other Writings*, Penguin Books, Harmondsworth, 1986.

Frazer, J. G., *Balder the Beautiful: The Fire-Festivals of Europe and the Doctrine of the External Soul*, 2 vols., Macmillan, London, 1930a.

Myths of the Origin of Fire, Macmillan, London, 1930b.

Freud, Sigmund, 'Group Psychology and the Analysis of the Ego' (translated from the German, 1921), Standard Edition 18, Hogarth Press, London, 1955, pp. 69–134.

'Civilization and Its Discontents' (translated from the German, 1930), Standard Edition 21, Hogarth Press, London, 1961, pp. 59–145.

'The Acquisition and Control of Fire' (translated from the German, 1932), Standard Edition 22, Hogarth Press, London, 1964a, pp. 185–93.

'New Introductory Lectures on Psychoanalysis' (translated from the German, 1933), Standard Edition 22, Hogarth Press, London, 1964b, pp. 3–182.

Freudenthal, Herbert, *Das Feuer im deutschen Glauben und Brauch*, De Gruyter, Berlin, 1931.

Frier, Bruce W., *Landlords and Tenants in Imperial Rome*, Princeton University Press, 1980.

Frisch, Max, *Biedermann und die Brandstifter*, Suhrkamp, Frankfurt, 1963.

Frost, L. E. and E. L. Jones, 'The Fire Gap and the Greater Durability of Nineteenth-century Cities', *Planning Perspectives* 4 (1989), pp. 333–47.

Furley, William D., *Studies in the Use of Fire in Ancient Greek Religion*, The Ayer Company, Salem, NH, 1981.

Fustel de Coulanges, Numa Denis, *The Ancient City* (translated from the French, 1864), Doubleday Anchor, New York, 1956.

Gales, B. P. A. and J. L. J. M. van Gerwen, *Sporen van leven en schade. Een geschiedenis en bronnenoverzicht van het Nederlandse verzekeringswezen*, NEHA, Amsterdam, 1988.

Geertz, Clifford, *Agricultural Involution: The Process of Ecological Change*, University of California Press, Berkeley, 1966.

Gellner, Ernest, *Plough, Sword and Book*, Collins Harvill, London, 1988.

Gleichmann, Peter Reinhart, 'Nacht und Zivilisation', in Martin Baethge (ed.), *Soziologie: Entdeckungen im Alltäglichen*, Campus, Frankfurt, 1983, pp. 174–95.

Goodale, Jane C., *Tiwi Wives: A Study of the Women of Melville Island, North Australia*, University of Washington Press, Seattle, 1971.

Gottwald, Norman, *The Tribes of Yahweh: A Sociology of the Religion of Liberated Israel 1250–1050 BCE*, Orbis Books, Maryknoll, NY, 1979.

Goudsblom, Johan, *Sociology in the Balance* (translated from the Dutch, 1973), Basil Blackwell, Oxford, 1977.

Nihilism and Culture (translated from the Dutch, 1960), Basil Blackwell, Oxford, 1980.

'Zur Untersuchung von Zivilisationsprozessen', in Peter Gleichmann, Johan Goudsblom and Hermann Korte (eds.), *Macht und Zivilisation*, Suhrkamp, Frankfurt, 1984, pp. 83–104.

'The Impact of the Domestication of Fire upon the Balance of Power between Human Groups and Other Animals', *Focaal* 13 (1990a), pp. 55–65.

'The Humanities and the Social Sciences', in E. Zürcher and T. Langendorff (eds.), *The Humanities in the Nineties: A View from the Netherlands*, Swets & Zeitlinger, Amsterdam, 1990b.

Goudsblom, Johan, E. L. Jones and Stephen Mennell, *Human History and Social Process*, Exeter University Press, 1989.

Gowlett, John, *Ascent to Civilization: The Archaeology of Early Man*, Knopf, London, 1984.

Gowlett, J. A. J., J. W. K. Harris, D. Walton and B. A. Wood, 'Early Archaeological Sites, Hominid Remains and Traces of Fire from Chesowanja, Kenya', *Nature* 294 (1981), pp. 125–9.

'Reply to Glynn Isaac', *Nature* 296 (1982), p. 870.

Graves, Robert, *The Greek Myths*, 2 vols., Penguin Books, Harmondsworth, 1955.

Graz, Louis, *Le feu dans 'l'Iliade' et 'l'Odyssée'*, Klincksieck, Paris, 1965.

Green-Hughes, E., *A History of Firefighting*, Moorland, Ashbourne, 1979.

Grinstein, Alexander, 'Stages in the Development of Control over Fire', *International Journal of Psychoanalysis* 33 (1952), pp. 416–20.

Gurney, O. R., *The Hittites* (2nd edn), Penguin Books, Harmondsworth, 1990.

Gutmann, Myron P., *Warfare and Rural Life in the Early Modern Low Countries*, Van Gorcum, Assen, 1980.

Hagger, Nicholas, *The Fire and the Stones: A Grand Unified Theory of World History and Religion*, Element Books, Longmead, Shaftesbury, 1991.

Haldon, J. F. and M. Byrne, 'A Possible Solution to the Problem of Greek Fire', *Byzantinische Zeitschrift* 70 (1977), pp. 91–9.

Hallam, Sylvia J., *Fire and Hearth: A Study of Aboriginal Usage and European Usurpation in South-western Australia*, Australian Institute of Aboriginal Studies, Canberra, 1975.

Hallpike, C. R., *The Principles of Social Evolution*, Clarendon Press, Oxford, 1986.

Hanson, Victor Davis, *Warfare and Agriculture in Classical Greece*, Giardini, Pisa, 1983.

Harden, Donald, *The Phoenicians* (2nd edn), Penguin Books, Harmondsworth, 1980.

Hardin, Garrett, *Filters against Folly*, Viking, New York, 1985.

Harris, Marvin, *Cannibals and Kings: The Origins of Cultures*, Random House, New York, 1977.

Hecht, Susanna and Alexander Cockburn, *The Fate of the Forest: Developers, Destroyers and Defenders of the Amazon*, 2nd edn, Penguin Books, Harmondsworth, 1990.

Hemming, John, *The Conquest of Peru* (2nd edn), Penguin Books, Harmondsworth, 1983.

Henley, Paul, *The Panare: Tradition and Change on the Amazonian Frontier*, Yale University Press, New Haven, 1982.

Hermansen, Gustav, *Ostia: Aspects of Roman City Life*, University of Alberta Press, Edmonton, 1981.

Herrin, Judith, *The Formation of Christendom*, Princeton University Press, 1987.

Hillel, Daniel J., *Out of the Earth: Civilization and the Life of the Soil*, Free Press, New York, 1991.

Hobsbawm, Eric and Terence Ranger (eds.), *The Invention of Tradition*, Cambridge University Press, 1983.

Hobsbawm, Eric and George Rudé, *Captain Swing*, Lawrence Wishart, London, 1969.

Hodder, Ian, *The Domestication of Europe. Structure and Contingency in Neolithic Societies*, Basil Blackwell, Oxford, 1990.

Hoffner, Harry Angier, Jr, *The Laws of the Hittites*, Ph.D. dissertation, Brandeis University, 1963.

Hohenberg, Paul M. and Lynn Hollen Lees, *The Making of Urban Europe 1000–1950*, Harvard University Press, Cambridge, Mass., 1985.

Hoover, Herbert Clark and Lou Henry, *Georgius Agricola: De Re Metallica* (translated from the Latin, 1556), Dover Publications, New York, 1950.

Hopkins, Keith, *Conquerors and Slaves. Sociological Studies in Roman History: 1,* Cambridge University Press, 1978.

  Death and Renewal. Sociological Studies in Roman History: 2, Cambridge University Press, 1983.

Horton, D. R., 'The Burning Question: Aborigines, Fire and Australian Ecosystems', *Mankind* 13 (1982), pp. 237–51.

Hoskins, W. G., *The Making of the English Landscape* (2nd edn), Penguin Books, Harmondsworth, 1970.

Hough, Walter, 'The Distribution of Man in Relation to the Invention of Fire-making Methods', *American Anthropologist* (n.s.) 18 (1916), pp. 257–63.

  Fire as an Agent in Human Culture, Bulletin 139, Smithsonian Institute, United States National Museum, Government Printing Office, Washington, 1926.

Howell, F. Clark, *Early Man*, Time–Life Books, New York, 1965.

  'Observations on the Earlier Phases of the European Paleolithic', in J. Desmond Clark and F. Clark Howell (eds.), *Recent Studies in Paleoanthropology, The American Anthropologist* 68 (1966), Special Issue, pp. 88–201.

Hudson, Charles M., *The Southeastern Indians*, University of Tennessee Press, Knoxville, 1976.

Hughes, J. D. F. and J. V. Thirgood, 'Deforestation in Ancient Greece and Rome: A Cause of Collapse?', *The Ecologist* 12 (1982), pp. 196–208.

Isaac, Glynn, 'Early Hominids and Fire at Chesowanja, Kenya', *Nature* 296 (1982), p. 870.

James, Steven R., 'Hominid Use of Fire in the Lower and Upper Pleistocene', *Current Anthopology* 30 (1989), pp. 1–26.

Johanson, Donald C. and Maitland A. Edey, *Lucy: The Beginnings of Humankind*, Warner Books, New York, 1982.

Johanson, Donald and James Shreeve, *Lucy's Child: The Discovery of a Human Ancestor*, William Morrow and Company, New York, 1989.

Jones, David, *Crime, Protest, Community and Police in Nineteenth-century Britain*, Routledge & Kegan Paul, London, 1982.

Jones, E. L., *The European Miracle: Environments, Economies, and Geopolitics in the History of Europe and Asia* (2nd edn), Cambridge University Press, 1987.

  Growth Recurring: Economic Change in World History, Clarendon Press, Oxford, 1988.

Jones, E. L., S. Porter and M. Turner, 'A Gazetteer of English Urban Fire Disasters', *Historical Geography Series: 13,* Geo Books, Norwich, 1984.

Jones, Rhys, 'Fire-stick Farming', *Australian Natural History* 16 (1969), pp. 224–8.

  'East of Wallace's Line: Issues and Problems in the Colonization of the Australian Continent', in Mellars and Stringer, 1989, pp. 743–60.

Joseph, Edward D., 'Cremation, Fire, and Oral Aggression', *Psychoanalytic Quarterly* 8 (1960), pp. 98–104.

Kafry, Ditsa, 'Playing with Matches: Children and Fires', in Canter 1990, pp. 47–62.

Kirby, Jon P., 'Bush Fires and the Domestication of the Wild in Northern Ghana', *Culture and Development Series: 1*, Institute of Cross-cultural Studies, Tamale, 1987, pp. 14–30.

Kitching, C. J., 'Fire Disasters and Fire Relief in Sixteenth-century England: The Nantwich Fire of 1583', *Bulletin of the Institute of Historical Research* 54 (1981), pp. 171–87.

Klaatsch, Hermann, *Der Werdegang der Menschheit und die Entstehung der Kultur*, Bong, Berlin, 1920.

Kolata, Gina, 'Fire! New Ways to Prevent It', *Science* 235 (1987), pp. 281–2.

Komarek, E. V., Sr, 'Fire and the Ecology of Man', *Proceedings Sixth Annual Tall Timbers Fire Ecology Conference*, Tall Timbers Research Station, Tallahassee, Fa., 1967.

Komarek, E. V., 'Effects of Fire on Temperate Forests and Related Ecosystems: Southeastern United States', in Kozlowski and Ahlgren, 1974, pp. 251–78.

Konner, Melvin, *The Tangled Wing: Biological Constraints on the Human Spirit*, Holt, Rinehart & Winston, New York, 1982.

Konvitz, Joseph, *The Urban Millennium: The City-building Process from the Early Middle Ages to the Present*, Southern Illinois University Press, Carbondale, 1985.

Korem, Albin, *Bush Fire and Agricultural Development in Ghana*, Ghana Publishing Corporation, Tema, 1985.

Kortlandt, Adriaan, *New Perspectives on Ape and Human Evolution*, Stichting voor Psychobiologie, Amsterdam, 1972.

Kortlandt, A. and M. Kooij, 'Protohominid Behaviour in Primates', *Symposium of the Zoological Society of London* 10 (1963), pp. 61–88.

Kozlowski, T. T. and C. E. Ahlgren (eds.), *Fire and Ecosystems*, Academic Press, New York, 1974.

Kranendonk, Willem, *Society as Process: A Bibliography of Figurational Sociology in the Netherlands*, Sociologisch Instituut, Amsterdam, 1990.

Kuhnholtz-Lordat, G., *La terre incendiée: Essai d'agronomie comparée*, Editions de la Maison Carrée, Nimes, 1938.

Laing, Alastair, *Lighting*, Victoria and Albert Museum, London, 1982.

Lane, Frederic C., *Venice: A Maritime Republic*, Johns Hopkins University Press, Baltimore, 1973.

Latham, Robert and William Matthews (eds.), *The Diary of Samuel Pepys: 7* (1666), Bell, London, 1972.

Le Goff, Jacques, *The Birth of the Purgatory* (translated from the French, 1981), University of Chicago Press, 1984.

Lee, Richard B., *The !Kung San: Men, Women and Work in a Foraging Society*, Cambridge University Press, London, 1979.

Lee, Richard B. and I. De Vore (eds.), *Man the Hunter*, Aldine, Chicago, 1968.

Leibowitz, Leila, 'In the Beginning ... The Origins of the Sexual Division of Labour and the Development of the First Human Societies', in Stephanie Coontz and Peta Henderson (eds.), *Women's Work, Men's Property: The Origins of Gender and Class*, Verso Editions, London, 1985, pp. 43–75.

Lemche, N. P., *Early Israel: Anthropological and Historical Studies of the Israelite Society Before the Monarchy*, Brill, Leiden, 1985.

   Ancient Israel: A New History of Israelite Society, JSOT Press, Sheffield, 1988.

Lenski, Gerhard, Jean Lenski and Patrick Nolan, *Human Societies: An Introduction to Macrosociology* (6th edn), McGraw-Hill, New York, 1991.

Lévi-Strauss, Claude, *The Raw and the Cooked* (translated from the French, 1964), Harper & Row, New York, 1969.

   From Honey to Ashes (translated from the French, 1966), Harper & Row, New York, 1972.

The Elementary Structures of Kinship (translated from the French, 1949, 2nd edn), Beacon Press, Boston, 1976.

Lewis, Henry T., 'The Role of Fire in the Domestication of Plants and Animals in Southwest Asia: A Hypothesis', *Man* (n.s.) 7 (1972), pp. 195–222.

'Ecological and Technological Knowledge of Fire: Aborigines Versus Park Rangers in Northern Australia', *American Anthropologist* 91 (1989), pp. 940–61.

Lewis, Nolan D. C. and Helen Yarnell, *Pathological Firesetting (Pyromania)*, Nervous and Mental Disease Monographs, New York, 1951.

Liebermann, Philip, review of *Quest for Fire*, directed by Jean-Jacques Annaud, *American Anthropologist* 84 (1982), pp. 991–2.

Lloyd, Seton, *The Archaeology of Mesopotamia* (2nd edn), Thames & Hudson, London, 1984.

Longmate, Norman, *The Bombers: The RAF Offensive against Germany 1939–1945*, Hutchinson, London, 1983.

Lumsden, Charles J. and Edward O. Wilson, *Promethean Fire. Reflections on the Origin of Mind*, Harvard University Press, Cambridge, Mass. 1983.

Lyons, John W., *Fire*, Scientific American Library, New York, 1985.

McCloy, Shelby T., *Government Assistance in Eighteenth-century France*, Duke University Press, Durham, 1946.

McEvedy, Colin and Richard Jones, *Atlas of World Population History*, Penguin Books, Harmondsworth, 1978.

McGrew, W. C., 'Comment', *Current Anthropology* 30 (1989), pp. 16–17.

'Chimpanzee Material Culture: What are Its Limits and Why?', *Anthropology* in R. A. Foley (ed.), *The Origins of Animal Behaviour*, Unwin Hyman, London, 1991, pp. 13–22.

MacMullen, Ramsey, *Roman Social Relations: 50 BC to AD 284*, Yale University Press, New Haven, 1974.

McNeill, William H., *Plagues and Peoples*, Doubleday, Garden City, NY, 1976.

The Pursuit of Power: Technology, Armed Force, and Society since AD 1000, University of Chicago Press, 1982.

'Control and Catastrophe in Human Affairs', *Daedalus* 118, 1 (1989), pp. 1–12.

Maddin, Robert (ed.), *The Beginning of the Use of Metals and Alloys*, MIT Press, Cambridge, Mass., 1988.

Magnusson, Magnus and Hermann Pálsson (eds.), *Njall's Saga*, Penguin Books, Harmondsworth, 1960.

Malinowski, Bronislaw, *A Diary in the Strict Sense of the Term*, Athlone Press, London, 1967.

Mallove, Eugene F., *Fire from Ice: Searching for the Truth Behind the Cold Fusion Furor*, Wiley, New York, 1991

Mann, Michael, *The Sources of Power. 1. A History of Power from the Beginning to AD 1760*, Cambridge University Press, 1986.

Maple, J. H. C. (ed.), *The Technological Impact of JET on European Industry*, JET Joint Undertaking, Abingdon, Oxon., 1987.

Marshall, Lorna, *The !Kung of Nyae Nyae*, Harvard University Press, Cambridge, Mass., 1976.

Mathias, Peter, *The First Industrial Nation: An Economic History of Britain 1700–1814* (2nd edn), Methuen, London, 1983.

Medvedev, Zhores, *The Legacy of Chernobyl*, Norton, New York, 1990.

Meiggs, Russell, *Trees and Timber in the Ancient Mediterranean World*, Clarendon Press, Oxford, 1982.

Melbin, Murray, *Night as Frontier. Colonizing the World After Dark*, Free Press, New York, 1987.

Mellars, P. A., 'Fire Ecology, Animal Populations and Man: A Study of Some Ecological Relationships in Prehistory', *Proceedings of the Prehistoric Society* 42 (1976), pp. 15–45.

Mellars, Paul and Chris Stringer (eds.), *The Human Revolution: Behavioural and Biological Perspectives on the Origins of Modern Humans*, Edinburgh University Press, 1989.

Meyer, G. M. de and E. W. F. van den Elzen, *De verstening van Deventer: Huizen en mensen in de 14e eeuw*, Wolters-Noordhoff, Groningen, 1982.

Miller, Madeleine, J. Maxwell and John H. Hayes, *A History of Ancient Israel and Judah*, Westminster Press, Philadelphia, 1986.

Milne, Gustav, *The Great Fire of London*, Historical Publications, London, 1986.

Mitchell, George J., *World on Fire: Saving an Endangered Earth*, Scribner's, New York, 1991.

Mokri, Mohammad, *La lumière et le feu dans l'Iran ancien et leur démythification en Islam*, Editions Peeters, Leuven, 1982.

Mokyr, Joel, *The Lever of Riches: Technological Creativity and Economic Progress*, Oxford University Press, New York, 1990.

Moore, P. D., 'Fire: Catastrophic or Creative Force?', *Impact of Science on Society* 32 (1982), pp. 5–14.

Moore, R. I., *The Formation of a Persecuting Society: Power and Deviance in Western Europe, 950–1250*, Basil Blackwell, Oxford, 1987.

Morgenstern, Julian, *The Fire upon the Altar*, Quadrangle Books, Chicago, 1963.

Morris, Desmond, *The Naked Ape*, Jonathan Cape, London, 1967.

Muhly, James D., 'The Beginnings of Metallurgy in the Old World', in Maddin 1988, pp. 2–20.

Muir, J. V., 'Religion and the New Education: The Challenge of the Sophists', in P. E. Easterling and J. V. Muir (eds.), *Greek Religion and Society*, Cambridge University Press, 1985, pp. 191–230.

Müller, Klaus, *Geschichte der antiken Ethnographie und ethnologischen Theoriebildung: 1*, Franz Steiner, Wiesbaden, 1972.

Multatuli, *Max Havelaar, or the Coffee Auctions of the Dutch Trading Company* (translated from the Dutch, 1867, 2nd edn), University of Massachusetts Press, Amherst, 1982.

Mumford, Lewis, *The City in History*, Harcourt, New York, 1961.

Murdock, George Peter, 'The Current Status of the World's Hunting and Gathering Peoples', in Lee and De Vore 1968, pp. 13–20.

Myers, Norman, *The Primary Source: Tropical Forests and Our Future*, Norton, New York, 1984.

Naveh, Z., 'Effects of Fire in the Mediterranean Region', in Kozlowski and Ahlgren 1974, pp. 401–34.

Needham, Joseph, *Gunpowder as the Fourth Power, East and West*, Hong Kong University Press, 1985.

Neufeld, E., *The Hittite Laws*, Luzac & Co., London, 1951.

Newbold, R. F., 'Some Social and Economic Consequences of the AD 64 Fire at Rome', *Latomus* 33 (1974), pp. 858–69.

Ney, Tara and Anthony Gale (eds.), *Smoking and Human Behavior*, Wiley, New York, 1989.

Oakley, Kenneth, 'Fire as Palaeolithic Tool and Weapon', *Proceedings of the Prehistoric Society* 21 (1955), pp. 36–48.

O'Dea, William T., *The Social History of Lighting*, Routledge & Kegan Paul, London, 1958.

Oman, Charles, *A History of War in the Middle Ages. 2: 1278–1485* (2nd edn), Methuen, London, 1926.

Partington, J. R., *A History of Greek Fire and Gunpowder*, Heffer, Cambridge, 1960.

Perlès, Catherine, *Préhistoire du feu*, Masson, Paris, 1977.

'Les origines de la cuisine: L'acte alimentaire dans l'histoire de l'homme', *Communications* 31 (1979), pp. 4–14.

'Hearth and Home in the Old Stone Age', *Natural History* 90 (1981), pp. 38–41.

'La guerre du feu a-t-elle eu lieu?' Interview with Annick Miquel, *La Recherche* 13 (1982), pp. 390–91.

Perlin, John, *A Forest Journey: The Role of Wood in the Development of Civilization*, Norton, New York, 1989.

Perrin, Noel, *Giving Up the Gun: Japan's Reversion to the Sword 1543–1879*, Godine, Boston, 1979.

Peters, Charles R. and Eileen M. O'Brien, 'On Hominid Diet before Fire', *Current Anthropology* 25 (1984), pp. 358–60.

Price, Roger, *The Modernization of Rural France*, St Martin's Press, New York, 1983.

Prigogine, Ilya and Isabelle Stengers, *Order out of Chaos: Man's New Dialogue with Nature*, Random House, New York, 1984.

Pritchard, James B. (ed.), *The Ancient Near East* (3rd edn), Princeton University Press, 1969.

Pyne, Stephen J., *Fire in America: A Cultural History of Wildland and Rural Fire*, Princeton University Press, 1982.

Quack, H. P. G., *Herinneringen: Uit de levensjaren van H. P. G. Quack, 1834–1914* (2nd edn), Van Kampen and Zoon, Amsterdam, 1915.

Radcliffe-Brown, A. R., *The Andaman Islanders*, Cambridge University Press, 1922.

Radzinowicz, Leon, *A History of English Criminal Law and Its Administration from 1750. 1: The Movement for Reform*, Stevens & Sons, London, 1948.

Rainbird, J. S., 'The Fire Stations of Imperial Rome', *Papers of the British School of Rome* 54 (1986), pp. 147–70.

Raumolin, Jussi, 'Special Issue on Swidden Cultivation', *Suomen Antropologi* 12 (1987), pp. 185–279.

Renfrew, Colin, *The Emergence of Civilization*, Methuen, London, 1972.

Before Civilization (2nd edn), Penguin Books, Harmondsworth, 1976.

Renfrew, Colin and Paul Bahn, *Archaeology: Theories, Methods and Practice*, Thames & Hudson, London, 1991.

Reynolds, P. K. Baillie, *The Vigiles of Imperial Rome*, Oxford University Press, London, 1926.

Ribeiro, Darcy, *The Civilizational Process*, Harper Torchbooks, New York, 1968.

Roberts, Neil, *The Holocene: An Environmental History*, Basil Blackwell, Oxford, 1989.

Robinson, Olivia, 'Fire Prevention in Rome', *Revue Internationale des droits de l'antiquité* (3e sér.) 24 (1977), pp. 377–88.

Rogerson, John and Philip Davies, *The Old Testament World*, Cambridge University Press, 1989.

Rosen, Christine Meisner, *The Limits of Power: Great Fires and the Process of City Growth in America*, Cambridge University Press, 1986.

Rowley-Conwy, P., 'Slash and Burn in the Temperate European Neolithic', in Roger Mercer (ed.), *Farming Practice in British Prehistory*, University of Edinburgh Press, 1981, pp. 85–96.

Rushdie, Salman, *Imaginary Homelands*, Granta Books, London, 1991.

Russell, Emily W. B., 'Indian-Set Fires in the Forests of the Northeastern United States', *Ecology* 64 (1983), pp. 78–88.

Sahlins, Marshall, *Stone Age Economics*, Aldine, Chicago, 1972.

Ste Croix, G. E. M. de, *The Class Struggle in the Ancient Greek World*, Duckworth, London, 1981.

Salomon, Ernst von, *The Outlaws* (translated from the German, 1930), Jonathan Cape, London, 1931.

Sauer, Carl O., *Agricultural Origins and Dispersals*, The American Geographical Society, New York, 1952.

Selected Essays 1963–1975, Turtle Island Foundation, Berkeley, 1981.

Schoffeleers, J. M., 'The Religious Significance of Bush Fires in Malawi', *Cahiers des Religions Africaines* 10 (1971), pp. 271–81.

'Introduction', in J. M. Schoffeleers (ed.), *Guardians of the Land*, Mambo Press, Gwelo Zimbabwe, 1978, pp. 1–46.

Schulte, Regina, 'Feuer im Dorf', in Heinz Reif (ed.), *Räuber, Volk und Obrigkeit: Studien zur Geschichte der Kriminalität in Deutschland seit dem 18. Jahrhundert*, Suhrkamp, Frankfurt, 1984, pp. 100–52.

Shostak, Marjorie, *Nisa: The Life and Words of a !Kung Woman*, Vintage Books, New York, 1981.

Sicilia, David B., 'Steam Power and the Progress of Industry in the Late Nineteenth Century', *Theory and Society* 15 (1986), pp. 287–99.

Sigaut, François, *L'agriculture et le feu: Role et place du feu dans les techniques de préparation du champ de l'ancienne agriculture européenne*, Mouton, Paris, 1975.

Simmons, I. G., *Changing the Face of the Earth: Culture, Environment, History*, Basil Blackwell, Oxford, 1989.

Simons, L. M. R., *Flamma Aeterna: Studie over de betekenis van het vuur in de cultus van de Hellenistisch-Romeinse oudheid*, Jasonpers, Amsterdam, 1949.

Sjoberg, Gideon, *The Preindustrial City*, Free Press, New York, 1960.

Slicher van Bath, B. H., *The Agrarian History of Western Europe, AD 501–1850*, Edward Arnold, London, 1963.

Snodgrass, A. M., 'An Historical Homeric Society?', *Journal of Hellenic Studies* 94 (1974), pp. 114–25.

Spiegel, Shalom, *The Last Trial: On the Legends and Lore of the Command to Abraham to Offer Isaac as a Sacrifice: The Akedah* (translated from the Ivrit), Pantheon Books, New York, 1969.

Staal, Frits, *Agni: The Vedic Ritual of the Fire Altar*, 2 vols., University of California Press, Berkeley, 1983.

Stahl, Ann Brower, 'Hominid Dietary Selection before Fire', *Current Anthropology* 25 (1984), pp. 151–68.

Stavrianos, L. S., *Lifelines From Our Past: A New World History*, Pantheon Books, New York, 1990.

Steensberg, Axel, *Draved: An Experiment in Stone Age Agriculture – Burning, Sowing and Harvesting*, National Museum of Denmark, Copenhagen, 1979.

New Guinea Gardens: A Study of Husbandry with Parallels in Prehistoric Europe, Academic Press, London, 1980.

Steinen, Karl von den, *Unter den Naturvölkern Zentral-Brasiliens*, Dietrich Reimer, Berlin, 1894.

Stewart, Omer C., 'Fire as the First Great Force Employed by Man', in W. Thomas 1956, pp. 115–33.

Stone, John and Stephen Mennell, *Alexis de Tocqueville on Democracy, Revolution, and Society*, University of Chicago Press, 1980.

Stone, Leo, 'Remarks on Certain Unique Conditions of Human Aggression (the Hand, Speech, and the Use of Fire)', *Journal of the American Psychoanalytic Association* 27 (1979), pp. 27–63.

Sumption, Jonathan, *The Albigensian Crusade*, Faber & Faber, London, 1978.

Swaan, Abram de, *In Care of the State: Health Care, Education and Welfare in Europe and the USA in the Modern Era*, Polity Press, Oxford, 1988.

Talbot, Lee M., 'Man's Role in Managing the Global Environment', in Daniel B. Botkin *et al.* (eds.), *Changing the Global Environment: Perspectives on Human Involvement*, Academic Press, New York, 1989.

Te Brake, William H., 'Air Pollution and Fuel Crises in Pre-industrial London, 1250–1650', *Technology and Culture* 16 (1975), pp. 337–59.

Thomas, Keith, *Religion and the Decline of Magic*, Weidenfeld & Nicolson, London, 1971.

Man and the Natural World: Changing Attitudes in England 1500–1800, Allen Lane, London, 1983.

Thomas, William L., Jr (ed.), *Man's Role in Changing the Face of the Earth*, University of Chicago Press, 1956.

Thornton, Peter, *Authentic Decor: The Domestic Interior 1620–1920*, Weidenfeld & Nicolson, London, 1984.

Trigger, Bruce G., *The Children of Aataensic: A History of the Huron People. Part 1*, McGill-Queen's University Press, Montreal, 1976.

Trinder, B., *The Making of the Industrial Landscape*, Dent, London, 1982.

Tunzelmann, G. N. von, *Steam Power and British Industrialization to 1860*, Clarendon Press, Oxford, 1978.

Tylor, Edward Burnett, *Researches into the Early History of Mankind and the Development of Civilization* (2nd edn), John Murray, London, 1870.

Unger, Richard W., 'Energy Sources for the Dutch Golden Age: Peat, Wind, and Coal', *Research in Economic History* 9 (1984), pp. 221–53.

Van Creveld, Martin, *Technology and War: From 2000 BC to the Present*, Free Press, New York, 1991.

Vreeland, Robert G. and Bernard M. Levin, 'Psychological Aspects of Firesetting', in Canter 1990, pp. 31–46.

Vries, Lyckle de, *Jan van der Heyden*, Meulenhoff-Landshoff, Amsterdam, 1984.

Wallerstein, Immanuel, *The Modern World System. 3. The Second Era of Great Expansion of the Capitalist World Economy, 1730–1840s*, Academic Press, New York, 1989.

Wallington, Neil, *Images of Fire: 150 Years of Fire-fighting*, David & Charles, Newton Abbot, 1989.

Washburn, S. L. and C. S. Lancaster, 'The Evolution of Hunting', in S. L. Washburn and Phyllis C. Jay (eds.), *Perspectives on Human Evolution. I,* Holt, Rinehart & Winston, New York, 1968, pp. 213–29.

Weber, Egon, *Peasants into Frenchmen: The Modernization of Rural France 1870–1914*, Stanford University Press, 1976.

Weber, Max, *The Protestant Ethic and the Spirit of Capitalism* (translated from the German, 1905–6), Scribner, New York, 1930.

 Ancient Judaism (translated from the German, 1917–19), Free Press, Glencoe, Ill., 1952.

Werner, P., *De incendiis urbis Romae aetate imperatorum*, Diss., Leipzig, 1906.

Wertime, Theodore A. and Steven F. Wertime (eds.), *Early Pyrotechnology: The Evolution of the First Fire-using Industries*, Smithsonian Institute Press, Washington, DC, 1982.

West, M. L. (ed.), *Hesiod: Works and Days*, Clarendon Press, Oxford, 1978.

White, K. D., *Roman Farming*, Thames & Hudson, London, 1984.

Wieërs, Thieu, *Wij zullen u met assen lonen! De bokkerijders in het Maasland* (2nd edn), Ten Bos, Nieuwkerken, 1986.

Willems, H. G. M., 'Onrust in een interbellum: moordbranders in Overijssel 1529–1566', *Overijsselse historische bijdragen* 96 (1981), pp. 51–70.

Williams, Trevor I., *A Short History of Twentieth-century Technology*, Clarendon Press, Oxford, 1982.

Wolf, Eric R., *Sons of the Shaking Earth: The People of Mexico and Guatemala*, University of Chicago Press, 1959.

Wright, Gordon, *Between the Guillotine and Liberty: Two Centuries of the Crime Problem in France*, Oxford University Press, New York, 1983.

Wrigley, E. A., *People, Cities and Wealth: The Transformation of Traditional Society*, Basil Blackwell, Oxford, 1987.

 Continuity, Chance and Change: The Character of the Industrial Revolution in England, Cambridge University Press, 1988.

Yavetz, Z., 'The Living Conditions of the Urban Plebs in Republican Rome', *Latomus* 17 (1958), pp. 500–17.

Yergin, Daniel, *The Prize: The Epic Quest for Oil, Money and Power*, Simon and Schuster, New York, 1991.

Zeeuw, J. W. de, 'Peat and the Dutch Golden Age: The Historical Meaning of Energy-attainability', *AAG Bijdragen* 21 (Wageningen, 1978), pp. 3–31.

Zeitlin, Irving M., *Ancient Judaism: Biblical Criticism from Max Weber to the Present*, Polity Press, Cambridge, 1984.

Zimmerman, Carle C., *Family and Civilization*, Harper & Brothers, New York, 1947.

INDEX

═══

Aaron, 85
Aborigines, 32, 47
Abraham, 73, 74–6, 93
Achan, 88
Achilles, 100–101, 102
active use of fire, transition to, 16–20
Aeneid (Virgil), 100
Afghanistan, 184, 185
Africa, 17, 25, 28, 49, 154, 191, 197, 208
Agni, 131
agrarian destruction, 108
Agrarian History of Western Europe (Slicker van Bath), 154
agrarian regime, 45, 77–8, 80, 93, 103–4, 123, 130
agrarianization
    control of fire as prerequisite, 8, 44–7
    dominant trend, 47, 57, 60
    ecological transition, as, 42–4
    in Europe, 50
    increased productivity, 52–4
agriculture
    devastation of, 108
    intensification, 53–5, 57–8
    labour-intensive methods, 55, 57–8
    modern, 173
Ahab, 82
Ai, 88
air pollution, 152–3
air raids, 182–3, 226
Albigensians, 135
alchemy, 2, 161
Alexander the Great, 96, 108
Alexandria, 121

Algonquin, 36
Altamira, 61
altars
    Christian churches, in, 132
    domestic, 119
    see also sacrifices; temple fires
America
    American Indians: fire economy, 30–31; fire torture, 206; population, 42, 45; slash and burn by, 48; submission of, 143, 170
    USA: conflagrations, 176–7; energy consumption, 165, 207; growth of cities, 177; lynchings with fire, 208
Amos, 85
Amsterdam, 149, 178
Anabaptists, 159
Andaman Islands, 204–5
Angers, 155–6
animals
    birds carrying fire, 217
    cat massacre in Paris, 122, 133
    chimpanzees, 21–2, 24–5, 33–4
    domestication of, 43–4
    elephants, 26–7
    horses, 155
    human dominance over, 22–3, 28, 38, 42–3
    leopards, 101
    primates, limited ability with fire, 24–5, 33–4
    reaction to forest fires, 13–15
    sabre-toothed tiger, 26

    sacrifices, 122
    widening gap between humans and, 24–8
    wild, 45
Annals (Tacitus), 122
Annaud, Jean-Jacques, 10
anomie, 203
anthropology, 2–3, 189–92, 197–8
anxieties, 14, 20, 124, 136, 195–6
Apollo, 120
Arabs, 128, 137–8
archaeology, 4
Archimedes, 109
Aristotle, 113
arms race, 142–3
arson, 117–18, 148–9, 157–61, 181, 184–5, 201–3, 238
art, 111, 226
Artemis, 122
artisans, 162, *see also* craftspeople; working classes
Ash Wednesday fires, 192
ashes, 2, 49
Assyria, 73, 88, 90
Atar, 131
Athena, 120
Athens, 96, 106, 107, 109, 111, 126
Augustine, Saint, 97, 125, 129
Augustus, 114, 115
Australia, 31–2, 42, 45, 47, 170, 177, 197
Australopithecus prometheus, 17
autos-da-fé, 137

Baal, 77, 82–3, 91
Babylon, 65–8, 74

Bacchius, 111
Bachelard, Gaston, 41, 195–6
Baghdad, 131
Bali, 59
Baltimore, 181
Bede, Venerable, 56
Before Civilization (Renfrew), 4
Benedict, Ruth, 5
Berlin, 151
Binford, Lewis, 27
Birmingham, 177
Black Death, 153, 154
Blainey, Geoffrey, 190
Boerhaave, Herman, 162
Bosch, Hieronymus, 136
Boserup, Ester, 52–3
Bottéro, Jean, 65–6
Boyle, Robert, 143
Brahmins, 111
Brain, C.K., 23–4
Brandenburg, 151
branding, 68
Braudel, Fernand, 71, 144
Brazil, 187
Brewer, Stella, 34–5
brick, 144, 155, 177
'brickification', 145, 148
Brink, A.S., 25
building materials, 71, 144, 177
building regulations, 114, 144–6
Bulgaria, 138
Burford, Alison, 110–11
Burkert, Walter, 101–2
bush fires
    abandoned fires, from, 29
    cultural motives for, 189–91
Byzantium, 128, 130, 137–8, 140, 143

camp fires, 28–9
Canaan, 82, 95
Canada, 181
candles, 146, 155, 205–6, 210
cannon, 141–2
capital, 8, 78, 118, 145, 150
Carcopino, Jérome, 112
Carneiro, Robert, 187
Carthage, 97, 109, 185
catastrophes, vulnerability to, 44
Celsius, Anders, 162
central heating, 224

Changes in the Land (Cronon), 31
charcoal, 62, 66, 92, 93, 153
Chatwin, Bruce, 26, 101
Chernobyl, 213
Chesowanja, Kenya, 17
Chicago, 176–7, 181
children
    learning about fire, 196–8
    playing with fire, 200–201
    prohibitions on, 41, 196
chimneys, 152
chimpanzees, 21–2, 24–5, 33–4
China, 2, 57, 128, 141–2, 154
christendom, 122–5, 130–37
Churchill, Sir Winston, 184, 185
Cipolla, Carlo M., 143
cities, *see* hazards, urban
City of Gold (St Augustine), 97
civilization
    control of fire and, 3–6, 20, 41, 48–9, 97, 215
    definition, 3–6, 216
civilizing campaigns, 77, 124, 134, 146, 179–80, 199
civilizing constraints, 104, 142
Civilizing Process (Elias), 4–6, 128–9, 133, 206
civilizing processes, 6–8, 41, 48–9, 129–30, 193, 196, 210, 212
Clark, J.G.D., 50–51, 54
clearing land, 27, 28–33, 48, 186
clergy, 129
climatic changes, 30
coal, 152–3, 165, 168, 172
Cologne, 183
combustion process, 1, 33, 162, 173
compensation, 116–17, 150–52
conflagrations
    ancient Greece, 107
    industrial age, 176–82
    pre-industrial Europe, 144, 148–50
    Roman empire, 113, 115–18
    Russia, 116
    Third World, 114, 180
    wealth affecting, 206–8
Conklin, Harold C., 186
Constantine, 125
Constantine Porphyrogenitus, 138

Constantinople, 138
continuity, 40, 128
control and dependency, 10, 71, 168, 172, 174, 211
control of fire
    advances, 161, 170, 209–15
    consequences, 24, 37–8
    functions, 37–41
    increased, for humanity as a whole, 209–15
    individual acquisition of, 194–203
    pre-industrial Europe, 128–30
    social variations, 203–9
cooking, 13–14, 21–2, 23–5, 111–12
cosmology, 2, 162–3, 194
craftspeople, 58–9, 66, 93, 155, 161
cremation, 59, 100–102
Cronon, William, 31
cultural convergence, 9, 57, 94, 203
cultural divergence, 9, 57, 94, 106, 203
culture
    concept, 4–5, 216
curfew, 69, 146

Daniel, 87
Dart, Raymond, 16–17
Darwin, Charles, 2
David, 89
decivilizing processes, 106
deferred gratification, *see* detour behaviour
deforestation, 50, 92, 125–7, 141–5, 152–5, 188, 191–3
    see also forest fires; slash and burn
Delium, 107
Delphi, 120
dependency, *see* control and dependence
    on fire, 39–40
Descent of Man (Darwin), 2
destruction, 1, 11, 14, 33–4, 41, 44, 60, 71, 93, 139, 161, 182, 189, 202, 208
destructivity, 64, 182, 201–2
detour behaviour, 17, 45
Deuteronomy, 87–8
Deventer, Netherlands, 145
Dickens, Charles, 48, 166

differentiation in behaviour and power, 24, 26, 56–8, 170, 203–9
digestive system of humans, 34
Dinofelis, 26, 101
Diocletian, 125
Diodorus Cassius, 116
discipline, 22, 35, 41, 88, 94, 140, 149, 169, 179
distinction, 36, 206, 210
divine anger, fire as sign of, 85–7
divine power, fire as sign of, 80–84
division of labour, 40
domestic use of fire
  Greece and Rome, 111–12
  Israel, ancient, 92
  Mesopotamia, 66
  modern world, 200
domestication of fire
  accounts of, 2, 19, 97
  civilizing process, as, 6–8
  continued, 65, 78, 161, 195
  hominids, exclusive to, 16–18
  nature of, 33
  transition to, 16–20
domestication of plants and animals, 43–4, 50
dominant trends, 47, 55–8, 196
Donne, John, 162–3
Dresden, 183
Durkheim, Emile, 158, 203

earthenware, 60–61
Ecclesiastical History of the English People (Bede), 56
ecological consequences of fire use today, 211
ecological regimes, 38
ecological strategies, 49, 212
ecological transitions, 8–9
ecology, 3, 10, 30, 33, 52, 80, 190, 192, 211
Edison, Thomas, 210
Egnatius Rufus, 115
Egypt, 73, 88, 90, 130
Eleanor, Queen of England, 153
electricity, 173–4, 210
elements, theory of four, 162–3, 194
elephants, 26–7
Eliade, Mircea, 203–4

Elias, Norbert, 4, 7, 10, 87, 128–9, 133, 169, 196, 206, 210
Elijah, 82–4
elimination contests, 21, 86, 169
Emergence of Civilization (Renfrew), 4
emotional pleasure from fire, 29, 102
energy
  alternative sources, 212
  consumption, 165, 207, 211–13
  conversion of, 109
  fossil stocks, 153
  new sources, 172–6
energy, nuclear, 212–14
England, 30, 144, 148, 150, 152–3, 154, 160, 167
Ephesus, 113
Eskimos, 36, 39
ethology, 21–2
Europe, 9, 47–54, 128
exchange, 40, 99
execution by fire, 67–8, 88
extensive growth, 38, 42–3, 165, 189, 209, 212
Ezekiel, 86–7

Fahrenheit, Gabriel, 162
familiarity with fire, 55, 58, 66, 75, 91–4, 196–200
Fertile Crescent (Middle East), 45, 47
Festinger, Leon, 195
fiction, 202
figurations, 26, 58–60, 158
film, 202
Finland, 51
Finly, M.I., 99
fire
  active use of, 16–20, 22, 59
  ambiguity of word, 28, 65–6
  destructive effects, 33, 59, 92, 132
  ecology, 12–13
  fear of, 13–14, 93, 195–6
  Greek meanings, 99–100
  nature of, 28, 33, 162–3
  passive use of, 16
  pictograms for, 65
Fire and Ecosystems, 192

fire-arms, 140–43
fire brigades
  industrial age, 178–82, 199
  pre-industrial Europe, 148–50
  pumps, 179
  Roman empire, 115, 117–18, 146–7, 149
fire cults, 68, 131
fire economy, 30
fire-exclusion programmes, 192–3
fire festivals, 59, 132–4
fire gap, 177, 180
fire-growers, *see* pyrophytes
Fire in America (Pyne), 31
fire insurance, *see* insurance
fire prevention
  Babylon, 67
  caution taken for granted, 67
  Hattusa, 68–70
  industrial age, 178–82
  pre-industrial Europe, 143–8
  Roman world, 113–18
fire-protected zones, 52, 55, 58, 155, 182
fire regime, 32, 38–40, 166, 196, 200, 214
fire signals, 37, 107
fire worship, 131
firebreaks, 67, 144
firestick farming, 32, 47
firewood gathering as detour behaviour, 18
flame-throwers, 108
Flanders, 153
Fontana Economic History of Europe (Cipolla), 143
food
  preservation of, 61
  production, 52–4
Forbes, R.J., 121
forest fires
  animals' reaction to, 13
  benefits of, 14–16, 46
  classical authors, in, 103
  today, 185–93
  see also deforestation, fire economy, slash and burn
formidability, 11, 38, 44, 54, 170
Forni, Gaetano, 47
fortresses, 108–9
fossil fuels, 153, 164, 172, 211

France, 133, 135, 142, 148, 155, 160
Franklin, Benjamin, 156–7
Franks, 138
Frazer, Sir James, 2, 132–3, 190
Frederick I of Brandenburg, 151
Frederick William, 151
Freud, Sigmund, 19
Freudenthal, Herbert, 132, 134, 136, 156
Frost, L.E., 150
fuel, 59, 92, 97, 151–3, 207, 211
and deforestation, 125–7
fuel-intensive production, 171–2, 175, 186
functions, 37–8, 109
funeral rites, 59
furnaces, 86–7, 104, 125, 161
Fustel de Coulagnes, N.D., 112, 119

gas, 164, 172, 210
gathering
change to agrarianization, 42
dependence on, 43
Gehenna valley, 92–3
Gellner, Ernest, 60
genetic structure of humans, 194–5
Genghis Khan, 64
Germany, 134, 150–51, 183
Ghana, 189–91
glass-making, 161
Gleichmann, Peter, 175
Gomorrah, 85–6
Goodale, Jane, 197
Graz, Louis, 99
Greece (ancient)
agrarian regime, 103–4
city–states, 96, 106, 113
fire in religion and mythology, 97
fire prevention, 113
Hesiod, 103–4
Homer, 98–102
military organizations, 95–6
social stratification and fire use, 109–12
society, 95
sources, 36
temple fires, 120

wars, age of, 104–9
Greece (modern), 196
Greek fire, 137–40
guilds, 64
gunpowder, 140–43
empires, 142
Gurney, O.R., 64
Gutmann, Myron, 142

Hallam, Sylvia, 31–2
Hamburg, 177
Hammurabi, code of, 66–7, 147, 150
Hanno, 185
Hanson, Victor D., 108
Harris, Marvin, 55, 80
Hattusa, 68–70
hay heating, 156
hazards
rural, 70–71, 155–6
urban: Babylon, 66–8; fire-protected zones, 52; Hattusa, 68–70; industrial age, 176–82; pre-industrial Europe, 143–8, 155–7
Hazor, 89
hearths, 52, 55, 102, 103, 111, 205, 209–10
hearth cult, 119–20
heat, 37–9, 175, 206, 211
heating, 38–9, 66, 112, 126, 155, 174–5, 211, 223
Hector, 100, 102
hell-fire, 134–7
Hephaestus, 97, 110, 120
heretics, 134–6, 206
Hero of Alexandria, 109
Herodotus, 101, 105–6, 204
Herostratus, 113
Herrin, Judith, 130
Hesiod, 103–4, 111
Hestia, 97, 119–20
Hinduism, 59, 131
Hiroshima, 183
Hittites, 64
Hobbes, Thomas, 129–30
Holland, *see* Netherlands
Homer, 96, 97, 98–101
Homo erectus, 17, 25–7
Homo sapiens, 17, 28
Hong Kong, 181
Hosea, 85
hoses, 147, 149

Hough, Walter, 211
Howell, F. Clark, 26–7
Hughes, J. Donald, 125–6
human monopoly in controlling fire
dominance over other species, 21–2, 44–5
earliest evidence, 16–17, 203
emergence of, 21
exclusivity and universality of, 23
hunting
change to agrarianization, 42
communal, 35
dependence on, 43
fire as weapon, 27–8
Huygens, Christian, 141
hypocaust, 112

ice ages, 30, 42, 45–6
Iceland, 223
Iliad (Homer), 96, 98–101
incendiarism, *see* arson
incense, *see* altars
India, 2, 51, 57, 101, 128, 206
Indians, *see* American Indians
Indonesia, 184
Industrial Revolution, 164–6
industrialization, 9, 164–72, 209
industry
capitalism, 168–9
early sources of energy, 164–72
mass production, 170
new sources of energy, 172–6
insurance, 116, 118, 150–52, 158, 181
intensive growth, 38, 42–3, 71, 145, 161, 176
interdependence, 87, 174
International Encyclopedia of the Social Sciences, 3
International Thermonuclear Experimental Reactor (ITER), 214
iron, 63, 89, 165
irrigation, 52, 55
Isaac, 73, 75–6
Isaiah, 85–6, 91
Islam, 130, 136
Ismarus, 102

Israel
  New Testament sources,
    123–5
  Old Testament sources,
    72–5
Ivan the Terrible, 116

Jacob, 73
Japan, 142, 183
Jehoiakim, King, 92
Jeremiah, 79, 85
Jericho, 88
Jerusalem, 73, 90–91, 109
Jesus Christ, 123
Jezebel, 82
Joint European Torus (JET),
  214
Jones, David, 160
Jones, E.L., 150, 177–8, 180
Joshua 87–9, 92
Judah, 73, 85
judaism, 72, 130
  see also Israel; Old
    Testament sources
Julius Caesar, 184
Juvenal, 114, 117

Kafry, Ditsa, 200
Kidron valley, 92
Kirby, Jon, 189–90
Konner, Melvin, 27
Koran, 136
Korem, Albin, 191
Kortlandt, Adriaan, 22
Kristallnacht, 183
Kuikuru Indians, 187, 190, 192
!Kung bushmen, 28, 197
Kuwait, 185

Laing, Alistair, 210
lamps, 52, 55
Lancaster, C.S., 23
land use and productivity,
  52–4
Lane, Frederic, 146
Lascaux, 61
Lavoisier, Antoine, L., 163
Le Goff, Jacques, 143
Lenski, Gerhard, 63–4
Levi-Strauss, Claude, 2, 40,
  190
Lewis, Henry, T., 46–7
Lewis, Nolan, 201
Liège, 167

light, 37–9, 112, 175, 205,
  210–11
lighthouses, 121
lightning, 12, 156–7
livestock, 78
living standards, *see* intensive
  growth
London
  air pollution, 152
  conflagrations, 206–7
  fire of 1666, 144, 148, 150,
    176–7
loss of fire, 40
Louis XIV, 133
Lübeck, 183
Lucretius, 97
Lundström, Johan, 171
luxury, 206

McCloy, Shelby, 148
  McNeill, William, 10, 39,
    142
magic, 84, 111
Makapansgat, South Africa, 17
Malawi, 191
Malinowski, Bronislaw, 186
Manchester, 170, 177
Marcus Crassus, 115
Marshall, Lorna, 198
Marx, Karl, 8, 169
matches, 170–2, 194, 204–5
Matthew, Saint, 123
Medvedev, Zhores, 213
Meiggs, Russell, 126
Mellars, Paul, 46–7
Mesopotamia, 9, 65–8, 73, 146
metallurgy, 62–3, 89–90, 99,
  110–11, 161–2
micro-climates, 39
Miletus, 105
military-agrarian societies, 60,
  64, 73, 75, 95, 98, 109,
  128, 149
military regime, 99–102
mining industry, 155, 167
Mithraism, 122
Molech, 76–7, 93
monopoly formation, 20–23,
  89, 169, 171–2
Moore, Peter D., 12
Moore, R.I., 135
Morris, Desmond, 22–3
Moscow, 116
Moses, 77, 80–81, 84, 85, 87
motor car, 173

Muhly, James, 61–2
Multatuli (E.D. Dekker), 184
Mumford, Lewis, 182
Mycenae, 96, 101
Myers, Norman, 188–89
myths, 2, 12, 97, 133–4, 156
Myths of the Origin of Fire
  (Frazer), 2

Nagasaki, 183
Naked Ape, The (Morris),
  22–3
napalm, 185
naphtha (neftar), 84, 138
narcissism of small differences,
  130–31
natural sciences, 2, 35, 162–3,
  199, 212–14
Nebuchadnezzar, 87
necklace murders, 208
Needham, John, 141
Nehemiah, 84, 90–91
Neolithic era
  European, 48–51
  forest fires from natural
    causes, 12
  hunting preceding
    pastoralism, 46–7
  slash and burn farmers, 50
Nero, 116, 125
Netherlands, 145, 149–50, 153,
  154, 159, 178, 183
New England, 30
New Testament sources,
  123–5, 136
New York, 206–7
  blackout, 175
Newbold, R.F., 116
Newcastle, 152, 177
Nicomedia, 117–18, 149
Njall's Saga, 224
nobility, 56, 129
Normandy, 142
Normans, 138
nuclear fusion, 213–14

Odysseus, 97, 102
Odyssey (Homer), 56, 96,
  98–102
Oeconomicus (Xenophon), 104
oikos (Greek family
  household), 99
oil, 138, 146, 155, 164, 172–3
Old Testament sources, 72–5,
  87–91

ordeal by fire, 40
organization, 58, 62, 110, 125, 179, 211
Orléans, 135
ovens, 52, 55, 60–61

Palaeolithic era, 60, 101, 164, 195, 214
Palestine, 73, 92, 122
Paris, 133
pastoralism, 46
Pathological Firesetting (Lewis and Yarnell), 201
Patrai, 122
Patrocles, 100
Patterns of Culture (Benedict), 5
pauperism, 160
Pausanias, 121–2
peasant houses, 70–71, 155–6
peasants, 57, 59, 70, 142
peat, 153
Peloponnesian War, The (Thucydides), 106–8
Pepys, Samuel, 148
Pericles, 96
Perlès, Catherine, 19, 35, 195
Persepolis, 109
Persia, 57, 68, 104–6
Peru, 51, 57
phantoms, 220
Philadelphia, 157
Philip II, 96
Philippines, 186
Philistines, 89
phlogiston, 163
Phoenicians, 85
physiological changes cooking, from, 34–6
Pinamonti, Pietro, 136
Pisans, 138
Plagues and Peoples (McNeill), 10
plants
  cereal crops, 44
  domestication of, 43–4
  human control over, 42–3
  pyrophytes, 15, 194
Plataea, 106
Plato, 113, 120, 126
pleasure from fire, 40
Pliny the Elder, 97–8
Pliny the Younger, 117
ploughing, 51, 55
Plutarch, 115

population increase, *see* extensive growth
Port Moresby, New Guinea, 185–6
postconditions, 38, 43
potters, 59, 60–61, 110–11, 161
power, balance of, 22, 25
power through fire, 26, 40, 58, 185
prairies, 31
preconditions, 18–20, 43, 47
Prehistoric Europe (Clark), 50
priests, 57, 59, 68, 74, 78–9, 94, 130–31, 191
Primary Source (Myers), 188
primates, 21, 24–5, 33–4
printing, 161
productivity of land, 43–4, 52–4
Prometheus, 2, 97, 120
property, accumulation of, 63
prophets, 79, 123
protection, 43, 61, 64, 99
Protestant ethic, 104
prytaneum (Greek temple), 120
psychiatry, 113, 201
psychoanalysis, 19, 25–36, 198, 200
psychology, 10, 32, 198
  arsonists, of, 201–2
public order, 67, 147, 179
purgatory, 87, 143
purification, 132–3, 136, 202, 207
Pyne, Stephen, 31
pyromania, *see* arson
pyrophobia, 201
pyrophytes, 15, 194
pyrotechnology, 64, 109–10

Quack, H.P.G., 166
Quest for Fire (Annaud), 10

Radcliffe-Brown, A.R., 204–5
Radzinowicz, Leon, 157
rain forests, 188–9
Rainbird, J.S., 115
Réaumur, R.A.F. de, 162
regime, *see* agrarian, ecological, fire, military, religious regime
Reichstag fire, 1933, 183
religion, fire in
  Greece and Rome, 119–25
  Israel, ancient, 72–87, 94

pre-industrial Europe, 130–37; burning at the stake, 134–7, fire cults, 130–31, fire festivals, 132–4, heathen customs, 131–2, narcissism of small differences, 131, organized religion, 130
religious regime, 80, 130, 134
Renfrew, Colin, 4, 63
Rennes, 148–49
Revelation of Saint John, 124
Ribeiro, Darcy, 203
riots, 208
rites, 40–41
rock paintings, 42
Roman empire
  aristocracy, 97
  conflagrations in, 113–18; compensation, 116–17
  deforestation in, 125–7
  fire of AD 64, 115–6, 122, 144, 148
  fire prevention, 113–18; building regulations, 114; fire brigade (*familia publica* and *privata*), 115, 117–18; tenants' open fires, 114–15
  hypocaust, 112
  insulae (tenements), 111–12, 114, 116
  military power, 86–97
  social stratification and fire use, 111–12
  sources, 97
  temple fires, 121
Rotterdam, 183
Royal Society for the Protection of Life from Fire, 130
rubbish burning, 92–3, 132–3
Rudraprayag, India, 101
Rushdie, Salman, 206
Russia, 51, 116, 138, 213

sabre-toothed tiger, 26
sacking and burning, 64–5, 93, 183–4, 202
sacrifices, 74, 75–80, 119
  animal, 122
  human, 75–7
Sahlins, Marshall, 55
Saint John, Revelation, 124

Saint John's Day festival, 133–4
Salomom, Ernst von,, 183–4
Samson, 89
San Francisco earthquake, 1871, 176–7
Sanuto, Marino, 147
Sardis, 105
Sartre, Jean-Paul, 136
Sauer, Carl, 3, 30
Saul, 72
Schoffeleers, J.M., 191
science, 161–3
scorched earth, 183–4
sea coal, 152–3
secularization, 121
self-control, 11, 20, 201
self-domestication, 20
Serbia, 70
shifted cultivators, 189
shifting cultivation, *see* slash and burn
ship-burning, 100
Sierra Leone, 185
Simmons, I.G., 165
Simons, Lyda, 120
slash and burn
   environmental changes from, 51
   prevalence of, 47–51, 154, 185–93
   step in civilizing process, 48–9
slaves, 68, 70
Slicher van Bath, B.H., 154
Smith, Adam, 130
smiths, 59, 61–5, 89–90, 110–11, 131, 155
smoke, 1, 28, 29, 71, 152
smoking, 171, 199
social classes, 169–70, 205–6
social constraints, 129
social control, 10–11
social distinctions, 36–7
social fire, 195–6
social status, 145, 206, 223
social stratification, 57, 109–12, 206–7
sociology, 10, 24, 53, 75, 181
Sodom, 85–6
Solomon, 89
Spanish Civil War (1936–9), 182
Sparta, 96, 106

specialization, 58–9, 110, 121, 125, 181, 198, 211
species monopoly, 8, 20–4, 44, 214
Spencer, Herbert, 172
stages, 8, 17, 43, 47, 190, 196
Stahl, Anne Brower, 34
stakes, 134–7
state formation, 74, 88, 129
steam engines, 166–70
steam ships, 168
Steinen, Karl von den, 13–15
Sterkfontein, South Africa, 25
Stewart, Omer C., 3, 29
stone-splitting, 37
'stonification', 148, 147
Suetonius, 116
survival units, 38
Swartkrans, South Africa, 17
swidden agriculture, *see* slash and burn
symbolic burning, 207–8
symbolism, 100, 202, 207, 209
Syracuse, 109

Tacitus, 116, 118, 122, 148
Tarde, Gabriel, 160
Tasman, Abel, 31
tax collectors, 160
technology, 19, 109–10, 139, 142–3, 149, 161–3, 170, 179–80
temple fires, 69–70, 90–91, 109, 113, 120–21
theology, 131
thermodynamics, 163
thermometers, 162, 224
Third World, 114, 180, 188, 192
Thirgood, J.V., 125–6
Thirty Years War, 142
Thucydides, 106–8, 113
Tiwi, 197
Tocqueville, Alexis de, 129, 166, 170
Tokyo, 183
tool-sharpening, 37
Torralba, Spain, 26–7
torture, 206
towns, *see* hazards, urban
Trajan, 117, 118
trends, 57–8, 125, 174
triad of controls, 10, 19, 22
trickle-down effect, 135, 199

Trojans, 100–102
Tunzelmann, G.N. von, 167
Turkey, 138
Tylor, Edward B., 2
Tyros, 108

urban history, 223

vagrants, 158–60
Van Creveld, Martin, 108, 139
Van der Heyden, Jan, 147, 149, 178
vegetation, 30
Venice, 146
Versailles, 155
Vesta, 97, 121
Vietnam, 185
Vigiles, 115, 118
villages, *see* hazards, rural
violence, 60
Virgil, 100
Visigoths, 125
Vitruvius, 114
volcanic eruptions, 12
Voltaire, Arouet de, 225
Vulcanus, 97, 110

war and fire
   cremation of heroes, 100
   fuel consumption, 126–7
   Greece, ancient, 98–102, 104–9
   Greek fire, 137–40
   gunpowder, 140–43
   Israel, ancient, 87–91, 93–4
   modern war, 182–3
   sacking and burning, 64–5, 93, 183–4, 202
warmth, *see* heat
warriors, 56, 60, 61–5, 99, 142
Washburn, S.L., 23
water power, 200, 212
weapons, 63–5, 89
Weber, Max, 104
weighing instruments, 162–3
What is Sociology? (Elias), 87
White, K.D., 109
Whyte, Lynn, 103
Wickham-Jones, C.R., 62
widow-burning, 206
wind power, 200, 212
witch-burning, 68, 135, 206
working class, 169–70
Works and Days (Hesiod), 103–4, 111

Wrigley, A.E., 153, 164

Xenophon, 104

Yagaw Hanunóo, 186–7, 190, 192
Yarnell, Helen, 201

Zeus, 119, 120
Zhoukoudiem, Beijing, 17
Zoroastrianism, 59, 131